REAL
SCIENTISTS
DON'T WEAR TIES

REAL
SCIENTISTS
DON'T WEAR TIES
WHEN SCIENCE MEETS CULTURE

Sidney Perkowitz

JENNY STANFORD
PUBLISHING

Published by

Jenny Stanford Publishing Pte. Ltd.
Level 34, Centennial Tower
3 Temasek Avenue
Singapore 039190

Email: editorial@jennystanford.com
Web: www.jennystanford.com

British Library Cataloguing-in-Publication Data
A catalogue record for this book is available from the British Library.

Real Scientists Don't Wear Ties: When Science Meets Culture
Copyright © 2020 by Sidney Perkowitz

For photocopying of material in this volume, please pay a copying fee through the Copyright Clearance Center, Inc., 222 Rosewood Drive, Danvers, MA 01923, USA. In this case permission to photocopy is not required from the author.

Cover image: Sidney Perkowitz, courtesy Emory Photo Video/Ann Watson

ISBN 978-981-4800-68-6 (Hardcover)
ISBN 978-0-429-35145-7 (eBook)

To my beloved wife Sandy, Mike and Erica,
and Nora—I'm grateful that you're here.

Contents

Preface: Scientist, Writer, or Both?

I've wanted to be a scientist ever since I can remember, and when I became a successful research physicist, I was living my dream. But my early heroes also included writers—novelists such as F. Scott Fitzgerald, Sinclair Lewis, and J. D. Salinger, and science-fiction and fantasy writers such as Robert Heinlein, Ursula Le Guin, and J. R. R. Tolkien. Somehow the writing life appealed to me as much as science did. Of course, as a scientist, I had the opportunity and even the necessity to write journal articles that presented my research, producing over a hundred research pieces and several research-oriented books.

That was good training to express science directly and concisely, but though my research felt creative, presenting it in the rigid format of a scientific paper did not. And so when in the same year I reached two landmarks, my 50th birthday and the publication of my 100th research paper, I decided it was time for a different kind of writing. Though inspired by novels and imaginative fiction, I knew my strength would be to use my science background to inform and engage people who aren't scientists—to write popular science. That would also need plenty of imagination to find understandable examples and metaphors for abstract scientific ideas and to navigate the boundary between science and science fiction, which can illuminate science and where it is taking us.

My transition was helped by friends and colleagues at Emory University where I was Charles Howard Candler Professor of Physics, within Emory's commitment to good writing and interdisciplinary education. I got valuable support too from John Wilkes, founder and at the time director of the highly regarded science writing program at University of California, Santa Cruz, and from Peter Brown, then the editor of the regretfully long-gone magazine *The Sciences*.

My first science article for general readers appeared in the *Miami Herald* in 1989. From then until I retired from Emory as an emeritus professor in 2011, I carried on a two-track lifestyle: academic research and teaching, and writing pop science. Now I focus only on writing.

I hugely enjoyed lab research and now equally enjoy writing, which puts me into the blissful state psychologists call "flow." The hours fly by and I write more than I would have believed possible. When I selected items to put into this collection, besides ten books published or in progress, I could choose from over 160 pop science articles, short blog pieces to long-form essays. I've selected fifty that represent what I think is my best writing and that cover a variety of topics and a time span from early pieces until 2018.

My choices are organized into three categories that reflect current research, my own interests, and those, I hope, of non-scientists.

The first category, "Science," is about pure fundamental science, which aims to understand nature from the submicroscopic to the cosmic level. This desire motivated the ancient Greek natural philosophers and still drives researchers. In "Science" you'll find selections dealing with the big questions and theories of physics—relativity, quantum mechanics, and the nature of light. The second big area I cover is the study of the matter that makes up the world around us.

That last area is closely related to my own research. Many pop science books are written by theoretical physicists, who do not work in labs but use their own minds, math and experimental data gathered by others to conceive theories, the best known example being Einstein and relativity. I however was an experimentalist, working in my lab to study the properties of solids such as semiconductors and superconductors with lasers and other tools. That gives me a different view of science and links right to my second category, "Technology."

"Technology" covers the applications of pure science. Many appear in our daily lives and depend on the physical properties of materials such as the semiconductors in computer chips. Other uses involve biomedicine and social science. The pieces in this category range from current technology such as the laser and clinical medicine to science-fictional but maybe not impossible future technology like invisibility. They also present the growing human impact of new technology such as artificial intelligence (AI).

The last category, "Culture," stems from my belief that science can be found in every human activity and is an integral part of human culture. "Culture" covers science and science fiction in visual art, literature, film, and television; science in everyday life; and the culture of science itself in pieces that reveal how scientists think and behave (one of these, "Real Physicists Don't Wear Ties," about how scientists choose to dress, lent its title to this book. When I wrote it in 1991, there were far fewer women in science than today, so the traditional male accessory of the necktie has become even rarer among scientists).

I enjoyed selecting the variety of work I present here. I hope that readers of all backgrounds will find the pieces compelling as they have been written, with the science presented both correctly and understandably (and even with humor).

That's important because of an idea bigger than my own personal collection. In today's world where scientific fact often seems to receive less than its due, scientists owe it to themselves to convey science to the public for its benefit and for the benefit of science itself. If my experience as a scientist/writer inspires other scientists to express their own understanding of science, that would be a wonderful bonus.

Sidney Perkowitz
Atlanta, Georgia and Seattle, Washington, USA
Summer 2019

Acknowledgment

I am happy to acknowledge the efforts of Jenny Rompas and Stanford Chong of Jenny Stanford Publishing, who approached me about the possibility of a book. They liked my idea of an anthology of my writings and have proven a pleasure to work with along with their editorial team.

List of Illustrations

All images courtesy of the Everett Collection with additional credits as given:

Chapter 1

Science

Introduction

Fundamental science asks, "How does nature work?," a question we have only partly answered after millennia of effort and that continues to engage scientists. For me, "light" is an especially important part of nature. It was integral to my research, and through my long-time interest in visual art, I also know that light and vision greatly shape us as humans. Bringing all this together inspired my first articles and first popular science book, *Empire of Light: A History of Discovery In Science and Art* (1996).

The mysteries of light

This section begins with "Illuminating Light" (2013), an essay about light in science and culture written for the Vitra Design Museum in Germany. "True Colors" (1991) focuses on how and why we humans see the colors that we do; then "Light Dawns" (2015) describes the history of studying light and explains why it moves so fast.

Light is central to the theories that define modern physics: Einstein's relativity that makes the speed of light c, 300,000 kilometers per second, the universal speed limit, and uses c to define the spacetime fabric of the universe that produces gravitation; and

Real Scientists Don't Wear Ties: When Science Meets Culture
Sidney Perkowitz
Copyright © 2020 Sidney Perkowitz
ISBN 978-981-4800-68-6 (Hardcover), 978-0-429-35145-7 (eBook)
www.jennystanford.com

quantum mechanics, where the photon—the discrete unit of light energy Einstein discovered—is an essential elementary particle. Both theories contain mysteries: quantum mechanics because physicists do not understand why it works as well as it does and relativity because we are only beginning to grasp its predictions about black holes and the nature of time.

Quantum mechanics, relativity, strings, and time

"Nobody Knows the Quantum" (2017) illustrates that we really do not understand the quantum world, as shown by how hard it is to explain it. "Strange Devices" (1995) presents the weird aspects of quantum mechanics in more detail. They also appear in its applications, which have grown along with the rise of nanotechnology in developments such as quantum computing.

"These Georgia Tech Physicists Helped Prove Einstein Right" (2016) describes the Nobel Prize–winning Laser Interferometer Gravitational-Wave Observatory (LIGO) experiment that observed gravitational waves coming from two colliding black holes, confirming the last remaining prediction of general relativity. The piece also introduces two of the (still relatively rare) female physicists who helped make the discovery.

"Quantum Gravity" (2014) discusses the hope of combining quantum theory and relativity to form a grand Theory of Everything (ToE) for the entire universe and suggests that measuring quantum properties in space could greatly help. In 2018, an international research team proposed exactly that, so we may see progress toward a new theory of quantum gravity to replace string theory and lead to a ToE. "The Seductive Melody of the Strings" (1999), my review of *The Elegant Universe*—the book that gave broad exposure to string theory—points out that experimentally verifying this theory would be exceedingly difficult. Twenty years later, proof seems impossible and physicists are seeking other options for a theory of quantum gravity.

"Time Examined and Time Experienced" (2018) was chosen by *Physics World* magazine as among its best features of the year. The essay ponders the nature of time which along with space forms the pervasive fabric of reality and is woven into science and human existence; but its combined physical, cognitive, psychological, and emotional effects still elude our understanding.

Solids, liquids, gases, and more

To help explain the world around us, "The Six Elements: Visions of a Complex Universe" (2010) compares ancient ideas of the make-up of the cosmos to modern theories. "Stealth Science" (1992) presents solid-state physics, the quantum theory of solids, and "Froth with Meaning" (2000) examines the blend of gas and liquid that forms foams like the head on an espresso (see my book *Universal Foam 2.0: From Cappuccino to the Cosmos* (2015)). "Everything Worth Knowing About. . .Ice" (2017) from *Discover* Magazine's "Everything Worth Knowing" series focuses on ice as an unusual natural mineral and on its connections to the origin of water that is essential for life, and possible link to the origin of life itself.

Illuminating Light

Step outdoors on a clear night and gaze at any of the stars above. Depending on its distance from Earth, its light has traveled through space for years, centuries or millennia at enormous speed before it reaches your eyes to appear as a brilliant spark amid darkness. If, for some cosmic reason, the star were suddenly to brighten or grow dim, you would not see the change until that same long time span had gone by. Now step indoors and switch on a lamp. Its glow apparently illuminates the room in an instant, but that light must also travel from the lamp to your eyes. It does so at the same speed as light from a star and covers meters in mere nanoseconds. Your eyes and brain function more slowly than that, and so the room seemingly brightens or darkens instantaneously when you turn the lamp on or off.

Though the earthly and cosmic travel times are so different, they represent the same fundamental facts about light: it moves rapidly and in straight lines from a source to your eyes—or at least so it seemed to the early Greek philosophers who contemplated light. We have modified these conclusions as our knowledge of light's intangible nature has evolved into deeper understanding—deeper, but incomplete, for light still holds mysteries. Yet we know it well enough to predict its behavior and manipulate it with exquisite finesse for scientific research, technological application, and aesthetic use. Most important for us, perhaps, we also know how to create it to illuminate the night.

The prehistory of light

Although humanity's generation of artificial light dates back many thousands of years, light has been part of nature almost since the Universe was born. In the form of packets of energy called photons, light appeared soon after the Big Bang, the explosion of space, time, and energy that began the universe some 14 billion years ago. Now light continues as an essential part of the cosmos along with matter, which light helped to form. Einstein's equation $E = mc^2$ indicates that matter can change into energy, and energy into matter. The latter happened after the Big Bang when newborn photons collided and turned into electrons and other elementary particles. These eventually combined into today's stars, planets, and our own Earth, while the remaining light traversed space, as it still does. That primeval light, the cosmic background radiation, is an important clue to the Big Bang and the evolution of the universe.

Along with its role in the physical universe, light has also been essential for life. Light energy from the Sun helped create the complex molecules that came together to form life on our planet, and sunlight sustained life forms as they developed into plants and animals. As these became more complex and differentiated, they reacted to light in varied ways. Prehistoric humans surely responded with joy or fear to the Sun's rising and setting, respectively, without knowing the nature of its light. Perhaps as a result of those feelings, light has long played a role in human spiritual and divine beliefs and in the human desire to seek beauty and create art.

Critical experiments

For these reasons, when early Greek thinkers asked, "What is the nature of the world around us?," it was inevitable that they would also question the nature of light. That examination deepened as science developed after the Greeks. Later researchers closely examined light through measurements and experiments and theorized about it from those results. As scientific study answered old questions about light, it raised new ones that required further investigation until we have reached our present advanced but still incomplete understanding.

For example, the Greeks knew that light moves quickly or even at infinite speed from the simple observation that a distant mountain appears in one's vision apparently instantaneously as the eyes are opened, but they did not attempt to determine the speed. Centuries later, the great Renaissance scientist Galileo Galilei, the father of experimental science, proposed measuring the time it took a light ray to cover a known distance. Galileo may or may not have actually attempted this, but in any case the time intervals over earthly distances would have been too short to determine with contemporary timepieces.

Instead, the speed was first measured over astronomical dimensions through observations made by the Danish astronomer

Ole Roemer in 1676, at the Paris Observatory. He was determining how long it took Jupiter's moon Io to circle its planet, in hopes of establishing an astronomical clock for use by mariners. To his surprise, he found that the orbital period for Io seemed to depend on the month in which he took his data. In a critical insight, Roemer realized that the variation arose because light has a finite speed, and was delayed by different times as it traveled depending on the distance between Jupiter and our planet as they orbited the Sun.

Estimates of that distance, and the time intervals from Roemer's data, led to the first measured result for the speed of light, 210,000 kilometers per second—erroneous, but in the right range. Later measurements have refined this value until now it is one of the most precisely known parameters of nature, with a value of 299,792.458 kilometers per second (generally rounded to 300,000 kilometers per second). Even if the speed had been accurately determined then, that alone would not have revealed the essential nature of light, which became apparent only through other significant experiments.

In 1666, Isaac Newton made one of these momentous experiments when he sent a ray of sunlight through a triangular glass prism. The prism split the sun's colorless or "white" light into a rainbow, covering a gamut of colors that Newton described as red, orange, yellow, green, blue, indigo, and violet. To explain this spectrum, Newton drew on his deep understanding of mechanics to theorize that light is made of small particles or "corpuscles" that move in straight lines and are deflected differently by the prism according to their mechanical properties.

Conflicting theories and human vision

Newton's theory was taken as correct for over a century, but then another deceptively simple experiment suggested something different. Around 1801, the English physician and scientist Thomas Young sent light through two narrow slits cut in a thin sheet and saw an unexpected pattern projected onto a distant surface. Rather than two bright areas, one illuminated by each slit, Young observed alternating bright and dark bands across a wide area. This could not come from corpuscles streaming through the slits in straight lines but rather showed that light is an undulating wave. Like ocean waves encountering the edge of a seawall and spreading out in all directions, these light waves would spread out from each slit to meet or interfere beyond them, alternately reinforcing and negating each other to produce bright and dark areas.

But no one could then imagine what kind of wave a light wave might be. Over 60 years later, the Scottish theoretical physicist James Clerk Maxwell provided an answer. Remarkably, he had not set out to think about light but to derive equations that described all electric and magnetic phenomena. His result contained a surprise. It showed that an undulating combination of electric and magnetic fields—an electromagnetic wave—travels at exactly the speed of light. This suggested that light is electromagnetic in character, as was soon confirmed by experiment.

The electromagnetic wave nature of light explained its high speed, Young's experiment, and Newton's rainbow. Now we understood that the colors we see are related to certain electromagnetic wavelengths. The human eye senses a range of 750 to 400 nanometers, from red through orange and so on to blue and violet at the shortest wavelengths. White light contains all these wavelengths, each bent differently by a prism to produce a spectrum of wavelengths that we see as a rainbow of color.

This is not to say that electromagnetic waves completely explain how we perceive color, which also involves the physiology of the eye and how the brain interprets what the retina senses. The complexity of relating the physical nature of light to its perception became apparent with the publication of *Theory of Colours* (1810) by Johann Wolfgang von Goethe. The writer Goethe, who also had scientific interests, attempted to recreate Newton's experiments but instead mostly observed perceptual reactions to light. Though his conclusions about color did not overthrow Newton's observations, they later influenced artists and philosophers such as J. M. W. Turner and Ludwig Wittgenstein.

With electromagnetic theory, we now also understand that we see only a tiny part of a huge range of light filling the universe. Beyond the red lies the infrared with wavelengths longer than 750 nanometers, then microwaves and radio waves with wavelengths of millimeters to many meters. Ultraviolet light lies below 400 nanometers, followed by X-rays and gamma rays with wavelengths down to fractions of a nanometer. Each regime has the same electromagnetic nature but affects humans differently. Infrared light is felt as warmth on the skin, ultraviolet light tans the skin but can also damage living cells, and X-rays and gamma rays destroy cells.

Still, electromagnetic theory was not the last word about light. Our understanding again shifted because electromagnetic waves seemed to fail in explaining two critical phenomena: the photoelectric effect and the so-called "black-body" radiation from hot objects (so named for historical reasons). Exploring these failures in the late 19th and early 20th centuries led to quantum mechanics, the photon, and the latest understanding of light.

The rise of the photon

In the photoelectric effect, first seen in 1887, light shining on any of a certain group of metals was found to eject electrons from the metal. Somehow the energy of the light was transferred to an electron and broke it loose. That was puzzling, because it could happen only if light encountered the electron as if light were a particle, not a wave, like a billiard ball passing on its energy to a second billiard ball by colliding with it. The photoelectric effect seemed to imply the opposite of Young's experiment.

Another hint of something not fully understood came in that same era as scientists tried to answer a seemingly straightforward question: What are the properties of the light emitted by a hot body such as a flame, or the filament of the newly invented incandescent lamp, then first coming into wide use?

The basic mechanism for this black-body light was clear: high temperature makes the electrically charged atoms of a body vibrate, which creates electromagnetic waves according to Maxwell's theory. But when researchers calculated the properties of these waves, they obtained a worrisome result. It partly agreed with experiment, correctly showing how a body's temperature determines its long wavelength radiation; but whereas measurements show that a hot body emits less at short wavelengths, the theory predicted the opposite, that the radiation would increase toward infinity at ultraviolet and shorter wavelengths.

In 1900, the German physicist Max Planck resolved this "ultraviolet catastrophe" with a true breakthrough, the birth of quantum theory. By assuming that the atoms vibrated only at energies that were exact multiples of a certain basic unit—that is, in quantized steps—Planck modified the theory so that it perfectly predicted the measured black-body radiation at all wavelengths. The quantization of energy was a radical idea but its success in describing the light from a hot body could not be ignored.

Then, in 1905, Einstein quantized light itself. He explained the photoelectric effect by proposing that light is made of discrete packages of energy, later called "photons." Other evidence soon supported Einstein's insight, which earned him the 1921 Nobel Prize in physics. Photon theory also clarified why short-wavelength light like X-rays is more damaging than long-wavelength light, for the shorter the wavelength, the greater the photon energy. The theory was consistent with quantized atomic energies as well, for now it became clear that if an atom changed from one level to another, that would generate or absorb a photon.

Nevertheless, for all the success of the photon theory, the fact remains that light also undergoes wavelike interference such as Young observed. The unavoidable conclusion is that light can act as a wave or as a swarm of quantum objects, depending on how its properties are measured. This puzzling duality appears in the quantum physics of matter as well, where small particles like electrons can also behave like waves. Today, photons are part of the highly successful theory of light and matter called quantum electrodynamics (QED), which earned its creators the 1965 Nobel Prize in physics.

QED combines quantum theory with relativity, a reminder that Einstein's most famous theory also has implications for light. Electromagnetic theory tells us the speed of light but does not indicate that only light can reach this speed and that nothing can exceed it. That comes from Einstein's 1905 theory of special relativity, which shows that an object's mass grows as the object gains speed. The mass becomes infinite at the speed of light, which would require infinite energy to attain. But photons have zero mass and so only they can travel at light speed.

Einstein's general relativity, his theory of gravitation, shows another unexpected outcome, that gravity affects light so it does not necessarily follow straight lines. Newton had described gravity as an attractive force but in Einstein's more accurate theory, gravity arises when a massive body distorts the nearby spacetime continuum—the four-dimensional cosmic fabric made of three dimensions of space and one of time. A huge sun distorts this sufficiently to hold planets in their orbits, and to make light rays follow curved paths; indeed, general relativity was confirmed in 1919 by observing that light from a distant star was bent by our own sun.

Remarkable technology

Though the quantum and relativistic properties of light seem highly abstract, they have real technological uses. The relativistic properties of light are important for the global positioning system (GPS). A GPS device receives radio signals sent at the speed of light from space satellites in Earth orbits. The device calculates its position on the Earth's surface from the travel times of the radio waves, but these are affected by the Earth's gravitational influence on light. GPS software must include this to accurately calculate the location. In ordinary life, however, the effects are small enough to ignore, and light rays effectively travel in straight lines.

Quantum properties also enter into the modern technology of light. These include the quantization of energy in photons or at specific levels and other counterintuitive quantum features: an elementary particle like an electron or photon can represent several different physical states at once; two photons can be "entangled" or linked over large distances with no physical connection; and the properties of a photon can be transmitted through empty space to another distant photon, like the fantasy idea of teleportation.

One of the older quantum applications is the laser, which since its invention in 1960 has become vital in commerce, medicine, scientific research, and more. Huge lasers producing terawatts of power are used in attempts to induce nuclear fusion as a new clean energy source, and tiny lasers producing fractions of a watt send voice and data through the optical fiber network that spans our globe.

A laser depends on the fact that the electrons in each type of atom occupy a distinctive set of quantized energy levels, like rungs on a ladder. If a gas is placed in a tube with electrodes and a high voltage is applied, the electrons in the gas atoms are excited to higher rungs. Then they drop back to lower levels and give up their energy as photons with specific wavelengths, making the tube emit light. First explored in the late 19th century, such "discharge" sources are seen today in neon signs, named after the noble gas neon. This gas produces a red glow because its atoms yield photons with a wavelength of 633 nanometers; other gases generate other colors.

Such a source becomes a laser when a mirror is placed at each end of the tube. Photons from the gas atoms reflect back and forth between the mirrors and create more photons from the atoms they pass through a process called stimulated emission ("laser" stands for "light amplification by stimulated emission of radiation"). The resulting light beam is intense, tightly focused, and coherent (all its waves are in exact step) at a particular wavelength, such as red at 633 nanometers from neon or green at 514 nanometers from argon. Other laser media, which may be different gases or even solids, produce ultraviolet or infrared light.

The notion of multiple physical states, though hard to accept, is also basic to quantum physics. An ordinary object has definite values of its parameters such as momentum. But a quantum object like a photon or electron carries many physical possibilities, each of which becomes definite only when the property is measured. An electron, for instance, has a magnetic component whose north pole can point either up or down. A photon can be "polarized" so that the electric field associated with it points either vertically or horizontally. Until the electron actually passes through a device that measures its magnetism or the photon passes through a polarization filter, both possible values are valid.

This property, called superposition, allows photons to efficiently carry data in the form of binary bits with value zero or one, represented in digital devices by electrical switches that are either off or on. A photon polarized horizontally or vertically can also represent zero or one, but with superposition, *at the same time.* This is a huge gain, since a set of photonic quantum bits (or qubits) carries many times the data carried by the same number of ordinary bits. A computer with just 150 qubits, it is estimated, would match the combined power of all the world's most powerful conventional computers. This motivates the broad development of quantum computing with light that is now underway.

Entanglement and teleportation can manipulate data in new ways as well. Entanglement occurs when certain optical processes create a correlated pair of photons with an astonishing property: no matter how far apart, if you measure the polarization of one photon as either horizontal or vertical, the other immediately takes on the complementary vertical or horizontal value. There is no physical link, yet the second photon is somehow tied to the first through a "quantum channel." This also makes it possible to "teleport" the polarization state of a photon through empty space to another distant photon. Experiments confirm these effects over large distances and also show that they occur at multiples of the speed of light and perhaps even instantaneously, apparently violating the relativistic rule that nothing can travel faster than light.

These paradoxical effects are unquestionably real though we do not understand them. Nevertheless, the hidden quantum link allows data in the form of photon qubits to be transmitted in codes that cannot be broken or intercepted, a technique known as quantum cryptography. In this way the quantum properties of light guarantee Internet and telecommunications security for financial, industrial, and governmental operations.

The wave nature of light can also be exploited to produce remarkable technology, such as invisibility cloaking that hides an object from view in an almost magical manner. First achieved and announced to enormous interest in 2006, its basic idea can be grasped by imagining that you are standing in a flowing brook, trying to determine if there is a rock upstream in the water. If the moving water were to splash as it encounters the rock, the current would get distorted in a way that you could detect downstream. But if instead the water were to flow around the rock and then smoothly reform as if it had never been disturbed, the downstream current would show no sign of the obstruction.

An invisibility cloak does the same with light. It is a shield around an object that interacts with incoming light waves so that they leave along continuations of their original paths. An observer sees only apparently undisturbed light and the object is rendered invisible. Using the electromagnetic theory of light waves, researchers have built cloaking structures that successfully hide small objects illuminated by microwaves, infrared, or visible light. Extending this, we may someday see flexible person-sized cloaks as in the Harry Potter stories.

Defeating darkness

These 21st-century technologies contrast sharply with an old and familiar technology that is critical for humanity, the artificial illumination we take for granted whenever we turn on a lamp. The primitive roots of human-made light go too far back in prehistory to be determined, but must have originated with fire from a natural phenomenon such as lightning. It is unclear when fire first became a true tool for humans or their predecessors, but some limited evidence suggests that proto-humans used it as much as a million years ago.

We definitely know, though, that humans used fire for cooking 30,000 to 50,000 years ago and specifically for lighting at least 17,000 years ago. The Paleolithic people who painted vivid animal images on the walls of the Lascaux caves in southwestern France worked by the light of lamps found there. These were made from stones with shallow hollows that now contain traces of vegetable matter used to ignite animal fat. From these crude beginnings (though some lamps were artfully shaped and decorated), artificial lighting has developed along lines that followed and sometimes stimulated the science of light.

The first advances were candles, which date back to Roman times or earlier, and oil lamps. Both were employed in the Middle Ages and were later improved. Candles never became bright enough for comfortable nighttime reading but a smokeless type was invented in 1825. Oil lamps shed more light after the great French chemist Antoine Lavoisier showed that burning requires oxygen. In 1780, the Swiss scientist François Pierre Ami Argand designed a lamp with a hollow wick and a glass chimney. Rising warm air drew fresh oxygen though these to make a hotter flame that burned brighter and whiter. The reason why that was not known then, but now black-body theory shows that higher temperature produces more intense light with a color balance closer to the Sun's white light.

Other advances changed the nature of lighting from individual to systemic. First came the discovery of coal gas in the 18th century, a byproduct of coal processing that burned with a bright flame. By the 19th century, networks were being built to distribute this and other types of illuminating gas from central generators for uses such as street lighting, adopted in Paris in 1820, and domestic illumination.

This was more efficient than separate oil lamps but gas lighting still exhibited the drawbacks of open flame: unwanted heat, the danger of fire, and noxious byproducts that affected people's health and polluted their surroundings.

A step toward flameless illumination came in the early 1800s when the British scientist Humphry Davy invented a kind of electrical burning. Connecting a powerful electrical battery to two nearly touching carbon rods, he saw a spark jump or arc across the gap to heat the carbon, which glowed with a strong white light. Even into the 20th century, the carbon arc was useful for lighting movie sets, but its harsh light and constant need for replacement rods kept it from succeeding as a general light source.

Davy, however, had also noted that electric current passing through a piece of platinum heated it to a high temperature. This made it glow or incandesce but threatened to melt the platinum. The challenge was to develop a glowing filament that would not melt, or burn as it combined with oxygen. Improving on earlier inventions, by 1880, the American inventor Thomas Edison had used a filament of carbonized bamboo inside an evacuated glass bulb to create the first long-lived incandescent lamp, integrated into a complete lighting system. The modern version uses a tungsten filament, which has a high melting point, inside a glass globe in which the air has been replaced by inert argon and nitrogen.

Though incandescent lighting is the most widely used type of illumination, it is not the most desirable, for only a fraction of the electrical energy entering a light bulb emerges as visible light. As a hot black body at a temperature of several thousand degrees, the filament emits mostly infrared light that is felt as heat, with only a small portion in the visible region. It would be better and more efficient to generate visible light without heat or unwanted radiation from sources using quantum rather than thermal principles. As we have seen, quantized energy levels produce light at specific colors in discharge sources such as neon signs. But since we humans have evolved under sunlight, we strongly prefer its broad spectrum white light, so quantum lighting must be modified to provide acceptable general illumination.

One approach uses fluorescence, where materials called phosphors absorb short-wavelength light and re-emit it over a wide range of longer wavelengths. Edison and others experimented with fluorescent lighting in the 19th century but it achieved commercial success only in the 1950s. A fluorescent lamp consists of a glass tube filled with mercury vapor and excited with a voltage, which produces short wavelength ultraviolet light. That is absorbed by an opaque phosphor, coating the inside of the tube, which re-emits the light over a range of visible wavelengths to make the tube glow white.

The fluorescent process saves energy compared to incandescent lighting, but has its own drawbacks. Early fluorescent lamps produced harsh white light with more blue and less red than sunlight or incandescent light, which was unflattering to people's skin tones. Phosphors developed since then provide improved color balance, but many people still prefer "warmer" incandescent lighting. Also, mercury is toxic, so these lamps need careful handling and disposal.

Fortunately, fluorescent lamps are not the only quantum source. The newest type is the light-emitting diode (LED) made of semiconducting materials, the class of crystalline solids whose special electrical properties make them essential for digital electronics. Semiconductors can produce light through the process of electroluminescence, where a small voltage applied to a semiconductor structure called a diode raises its electrons to a higher energy level. When the electrons drop back to lower energy, they release photons at a wavelength determined by the properties of the semiconductor.

The first LEDs, invented in the early 1960s, produced only red light. Later other semiconducting compounds were developed to generate orange, yellow, and green. These early units were tiny dots that were useful only as indicators or readouts for devices like calculators. But further development in the mid-1990s led to blue LEDs as well as bigger and brighter units. Now it is possible to make white LEDs for general illumination, usually by converting blue light into white with a phosphor as in fluorescent lamps.

These quantum LED sources bring practical benefits. Because illumination consumes a large fraction of the world's electrical power, high efficiency LED lamps can reduce our dependence on fossil fuels without using toxic mercury. LEDs also add a new choice to the array of light sources available for decorative and aesthetic purposes, thus expanding the artistic impact of light.

Aesthetic light

Like natural light, artificial lighting strongly affects art and culture. Light from the stars awakened humanity's curiosity about the universe, and the spiritual power of sunlight has influenced how we worship. Likewise, artificial light has altered our sense of the world and our very civilization. By making nighttime less to be feared, artificial illumination has contributed to the growth of cities and public spaces. By allowing people to read at night, perhaps after a day's work that left no time for reading, it has led to a better-educated, more cultivated citizenry.

Within these influences, the fine and applied arts have also been changed and inspired by new light sources, often in unexpected or subtle ways. In the late 19th and early 20th centuries, stage actresses opposed the introduction of incandescent theatrical lighting, which they found less flattering than candlelight or gaslight. But the brighter incandescent sources showed actors more clearly to their audiences. That reduced the need for exaggerated gestures and facial expressions, leading to today's more naturalistic theater. Now versatile lighting produces a huge range of stage effects.

Although art has always been about light, modern visual artists and designers who embrace its science and technology can make light a true living presence in their works. These practitioners are quick to adopt new advances or to use ordinary sources in creative ways. Within a decade of the laser's invention in 1960, laser beams were appearing in works of art as well as enlivening action films and bringing new excitement to rock concerts. Lasers continue to provide spectacular visual effects, as in Hiro Yamagata's installation *Quantum Field X3* (2004) at the Guggenheim Museum in Bilbao, Spain.

On the other hand, also in the 1960s, Dan Flavin pioneered the artistic use of established sources by producing subtle auras of color from ordinary fluorescent lamps; James Turrell heightened the illusory quality of his works through the evenness of fluorescent lighting; and in *The True Artist Helps the World by Revealing Mystic Truths* (1967), Bruce Nauman expressed those words in a spiral of red and blue neon lighting, then went on to other works in neon.

The artistic use of both established and novel light sources is now common. The special qualities of fluorescent lamps remain important, for example, in Olafur Eliasson's experiments in the perception of light and space such as his permanent outdoor installation *Yellow Fog* (2008) in the center of Vienna. Other contemporary works depend on the latest technology. Jenny Holzer's *For SAAM* (2007), at the Smithsonian American Art Museum in Washington, D. C., is a cylindrical column over eight meters tall studded with white LEDs that display various texts swirling around the column. Leo Villareal's *Multiverse* (2008) surrounds a moving pedestrian walkway at the National Gallery of Art in Washington, D. C. with thousands of white LEDs that rapidly change to give viewers a heightened sense of motion amid hypnotically dazzling patterns. Neither of these works could have been created with older types of light sources.

The power of illumination

The ability to fully control artificial light is one outcome of the centuries of study that humanity has devoted to light, itself testimony to the enduring power of this intangible phenomenon. From its early beginnings, light remains pervasive throughout the universe, whether as the cosmic background radiation, the outpouring from our own Sun, or a central part of human technology.

As beings who see and analyze light, we recognize its importance in human existence and our perception of the world. Vision, the most highly developed of our senses, enlists much of the brain's cognitive power to make sense of the flood of energy and information that reaches us through our eyes, a flood that also carries spiritual and aesthetic meaning. Even this does not fully define the power of light, for it has given humanity the means to escape the night and grow as a civilized species.

This is why humanity has devoted so much thought and effort to making and understanding light. From the unimaginable blast of radiance that is a supernova star, to the comforting nighttime glow of a lamp, light illuminates us.

True Colors

Dinosaur skeletons with horrific teeth are what former New York City children remember from Sunday afternoons at the American Museum of Natural History; a treasure-house of natural wonders at the edge of Central Park. *Tyrannosaurus, Allosaurus,* and their menacing cousins could seize a child's imagination. They were portrayed as titans in a tumultuous world, preying on sluggish plant eaters while volcanoes—lurid in the museum dioramas—erupted in the distance. Not far from the dinosaur display, though, was a room with a quite different mood: a hall filled with colored minerals, crystals elegantly etched by nature, and gems faceted by human craftsmen. When my father and I were among the Sunday throngs, the austere but lustrous beauty of that display was what drew me in. In the geometry of the crystals was a reassuring serenity and order—an antidote to the turmoil of the dinosaurs' world. I did not know then that the shapes reflect a deeper regularity, at the very heart of solid matter.

The philosophers of ancient Greece, searching for unifying principles behind the diversity of the world, were the first to imagine that crystals, blades of grass, soil, and everything else might be made up of indivisible particles called atoms. The variety of atoms invoked was small, but combined in vast numbers, in proportions that varied for each solid, those few kinds of atoms were considered enough to explain the whole grand array of nature. The modern picture of matter grows out of these philosophical roots. Twentieth-century physics recognizes that atoms are not indivisible but, like Chinese boxes, have even smaller units inside them—protons, neutrons, and electrons. Chemistry explores how atoms join to form molecules, which make up a bewilderingly complex level of organization intermediate between the atoms and the familiar world of dinosaur skeletons and gems. Yet contemporary science still pictures the observed properties of matter as arising from simple units endlessly repeated in the subtle architecture of a solid.

Crystals give direct expression to this underlying regularity. Look closely at a pinch of table salt: the cubic shape of each grain is actually a large-scale portrait of a microscopic arrangement of sodium and chlorine atoms, the basic building blocks of the crystal.

Similarly, the sixfold filigree of a snowflake caught on your sleeve reflects the minute hexagonal packing of the water molecules. The crystal spangles on an old brass doorknob or a sheet of galvanized iron also bespeak an inner geometry: the boundaries between the visible grains mark changes in the orientation of the metal's crystal lattice.

Picking out different crystal forms takes study, even in a display of minerals and gems. What arrests the eye at once is the varied interplay of crystals and light. Why is a massive quartz crystal as clear as mountain air, though it is as dense as any rock? What accounts for the crimson stain of ruby, the blue of sapphire? Why do needlelike crystals of native silver glisten, reflecting light rather than transmitting it? Why do transparent crystals distort anything seen through them? Those effects too can be explained in terms of geometry, of the profound spatial order of each crystal. And even though many everyday materials are not crystalline, an understanding of how crystals interact with light goes a long way toward accounting for the look of the world as a whole. What science has learned in recent decades about the interaction of light and crystals has also opened the way to new technologies, making it possible to create materials that interact with light in new and useful ways.

The overwhelming number of particles in any sizable chunk of matter would seem to rule out any fundamental explanation of its properties. The copper in a handful of pennies contains enough electrons, if each corresponded to a human being, to populate a thousand trillion earths. But because of the limited variety of particles (electrons come in only one kind, and a given material is rarely made up of more than three or four kinds of atoms), there is a salvation. One can start with a single particle and go on to imagine how its properties might change as it encounters another, identical particle, and another, and so on. Thus one can build an understanding from the behavior of an individual particle to the quite different behavior of a crowd.

Several key properties of crystals can be appreciated if one starts with a single atom of hydrogen. Hydrogen is the simplest atom, made up of one proton and one electron. The classic image of this atom as a miniature solar system in which the lone electron orbits the proton is beguiling, but it ignores the complexities of quantum mechanics. They decree that the electron is not at all like a wee planet, with a well-defined position at every instant; instead it is spread out in space like a fog enveloping the nucleus.

Hydrogen has only one electron cloud, at its lowest energy spread spherically around the nucleus like a ball of fuzz. Other atoms have multiple electron clouds, some of which are far from spherical. Instead, they extend like tentacles in various directions from the nucleus. The tentacles, together with the interactions among the nuclei themselves, lock the atoms of a crystal into a rigid, three-dimensional structure. The oriented bonds between the atoms give the crystal a basic structural unit—a few atoms arranged in a cube, for example, or in a hexagon. Repeated over and over again to fill space, this geometric unit becomes a beat in a spatial rhythm, which is ultimately expressed in the overall crystal shape.

The electrons that help specify the crystal geometry are also responsible for the way the material responds when light strikes it. To picture the electronic structure of a crystal, return again to the single atom. According to quantum mechanics, a razor-sharp exactness compensates for the misty ambiguity of the electron clouds surrounding the atom: each electron can assume only one out of a set of discrete energies, rather like the rungs of a ladder. The electron must occupy one rung or another; the spaces between are off-limits. Ordinarily the electron stays on the lowest available rung—its lowest possible energy state. It can climb to a higher rung only if it is stimulated with precisely the amount of energy that corresponds to the space between the rungs. Too big or too small a step and it crashes back down, like a person on a real ladder.

Now suppose the lone atom, with its sharply defined energy levels, has neighbors less than a billionth of a meter away—the typical spacing of the atoms in a crystal lattice. The electron clouds of the adjacent atoms overlap; each electron "feels" the influence of its counterparts in the other atoms. Something odd now happens to the energy rungs. One might expect that each rung would make room for all the electrons to which its energy corresponds. But a decree of quantum mechanics, the exclusion principle, intervenes: no two electrons can have identical properties, including energy. To coexist they must assume slightly different energies. As electrons crowd in, each rung splits into closely spaced but distinct rungs, each accommodating one electron (or a maximum of two if they differ in the quantum-mechanical property called spin).

The splitting of the energy rungs continues as more atoms arrive; if a billion atoms are bound together in a crystal lattice, each energy rung fragments into as many as a billion closely spaced but separate rungs. Scale up the picture to the still larger number of atoms in any real crystal, and the rungs become so closely packed they form a continuous range—an energy band, in which an electron can be found at any energy you like.

Yet the process of energy splitting, which turns the energy levels of an isolated electron into energy bands, does not go on broadening the bands until they all merge into a continuum. The energy ladder remains a series of broad steps. Between successive steps are voids called band gaps, which exert a powerful influence over the crystal's response to light. Whereas the bands result from the interaction of the electrons with one another, the gaps are maintained by the interaction of the electrons with the three-dimensional architecture of the crystal itself.

To picture that interaction you need to call on another quantum-mechanical principle, which holds that an electron not only has no definite position in space but also has a dual identity, sometimes resembling a wave and sometimes a particle. If the microscopic world of a crystal lattice were governed by the laws of classical physics, which view the electron simply as a particle, moving electrons would ricochet through a crystal like hyperactive ping-pong balls. But electrons within the crystal lattice also resemble ocean waves surging under a pier, crashing and sloshing their way through a regular array of fixed pilings that correspond to the atoms in the lattice.

When waves move through a regular structure of this kind, many of them are wiped out by a phenomenon known as destructive interference. For example, ocean waves reflected by successive pilings will overlap with one another as well as with incoming waves. If the overlapping waves are out of phase—if the crest of one lines up with the trough of another—they cancel each other. Waves survive the interference only if their wavelengths are equal to the spacing between the pilings or can otherwise fit neatly into the structure. The periodic lattice of a crystal acts in the same way to filter the waves of the electrons, accommodating only those wavelengths that match its atomic geometry and excluding the rest. Because in quantum mechanics the wavelength of an electron is related to its energy, forbidden wavelengths are tantamount to forbidden energies, or band gaps.

Much of the splendid diversity of the visual world—the clarity of some substances, the shininess of others, the colors of many more—depends on the spacings of the energy bands and on their populations of electrons. The structure of the energy bands also determines how well a crystal can conduct electricity; in fact, the appearance of a crystal and its electrical properties are so closely related that it is worth describing the band structures of electrical insulators and how they differ from those of materials that conduct, whether poorly or well.

The energy bands of a crystal fill in from the bottom up. Some solids contain just enough electrons to fill their lower energy bands, leaving the higher bands empty. Such solids do not readily carry an electric current, because the electrons in a filled energy band are pinned down like marbles in a well-packed box, unable to move in response to an electric field. Whether the crystal can conduct electricity at all depends on how much energy it takes to promote an electron from the topmost filled band to the empty conduction band, the next band up in the hierarchy.

If the band gap is large, as it is in crystals such as quartz and diamond, electrons rarely reach the conduction band, and the material is a good insulator. If the gap is smaller, thermal energy— the incessant jostling of atoms in the crystal lattice—can be enough to kick some electrons into the conduction band. The crystal will then act as a conductor, albeit a very poor one. Such materials are called semiconductors, though the term is a relative one. At low temperatures the thermal motion of the atoms slows, and electrons stay entrenched in the lower bands; the material becomes a *de facto* insulator.

Other substances contain enough electrons to partly fill their topmost band. A minuscule amount of energy suffices to nudge electrons from the lower, filled part of the band to the empty region above. Once there, the electrons become highly mobile, able to roam freely through the solid. The partly occupied band acts as the conduction band, and the material readily carries current. Such a band structure is typical of a metal.

Thermal energy is not the only way of lofting an electron to the conduction band. Light also does it nicely. Photons, the minute quantum packages of light, deliver precise amounts of energy to the electrons in the lattice—more when the photon comes from the violet, or short-wavelength, end of the spectrum, and less when it comes from the red, or long-wavelength, end. A crystal exposed to white light is barraged by photons ranging in energy from 1.6 electron volts for the reddest of the visible photons to 3.3 electron volts for the deepest-violet photons. (An electron volt, a handy unit for minuscule energies, is the energy an electron gains in moving through an electric potential of 1 volt.)

If the band gap of an insulating material is wider than the most energetic photon of visible light, no electrons will be promoted across the band gap to the conduction band. The band gap of diamond, for instance, is 5.4 electron volts, and that of quartz is larger still. As a result light slips through the lattice without being absorbed; quartz and diamond are transparent, as are insulators in general.

Yet that picture can change dramatically if the perfection of the lattice is spoiled by a few atoms of an impurity or by defects such as missing or displaced atoms. Pure, insulating aluminum oxide is water-clear. In nature, though, aluminum oxide occurs in several gem forms, all of them brilliantly colored: ruby, blue sapphire, yellow sapphire. The striking colors of the gems result from foreign atoms scattered through the crystal lattices. Like isolated atoms, these impurities have their own energy levels, which differ from the energy bands of the host lattice.

The impurities thus add new steps to the energy structure, spaced so that some visible photons now carry enough energy to promote electrons. Defects, which locally change the geometry of the lattice, can also create new energy levels. Either way, some visible light can give up its energy to the lattice electrons and be absorbed.

In the case of ruby, scattered atoms of chromium introduce energy levels that enable the crystal to absorb violet, yellow, and green light. What passes through the gem is mainly red light with a hint of blue, which is why a good ruby has the purplish cast called pigeon blood. The impurities in blue and yellow sapphire—titanium and iron, respectively—create other sets of energy levels and distinctive patterns of absorption. The transparent mineral fluorite, on the other hand, can acquire color without any change in composition: it turns purple when bombarded with radiation, which displaces fluorine ions from their usual positions and warps the lattice geometry. New energy levels are created, capable of filtering specific energies out of the light; what escapes from the crystal looks purple.

Other than gems and minerals, the material world includes few textbook crystals. Glass is technically a liquid that has cooled below its freezing point without crystallizing; given enough time it can flow, as the distorted windows of old cathedrals attest. The molecules of a glass are oriented at random, in contrast to the rigid geometry of a crystal. Crystal structure can be found in most rocks, but it often takes the form of a multitude of crystalline grains that differ in composition and orientation. Plastics and the surface layers of living things are made up of giant molecules gathered in patterns far more complex than those of a crystal. And yet the insights gained from crystals into the relation between electronic structure and optical behavior go a long way toward explaining how the rest of the world looks.

The transparency of glass correlates with its excellence as an electrical insulator, and the scintillating colors of stained glass again testify to light-absorbing energy levels introduced by impurity atoms. Similarly, selective absorption by pigments accounts for the colors of dyes and printer's inks, and also for the dominant color of the natural world. Molecules of chlorophyll *a*, the natural pigment that collects solar energy to drive photosynthesis in the leaves of most plants, have electronic energy levels that absorb at the red and blue ends of the spectrum; only the intermediate colors the eye sees as green can escape when light plays over foliage.

In order to explain other aspects of the interplay of light and materials—the refraction, or bending, of light as it passes through an insulator, for example, and the luster of metals—the picture of energy bands in a crystal has to be supplemented. One must take a closer look at what takes place when light strikes the electrons of a crystalline solid, whether they are bound in the lattice of an insulator or semiconductor or drifting freely in a metal.

Think of each electron in an insulator or a semiconductor as a minute oscillator, rather like the string of a musical instrument or a child's swing. Both string and swing have a natural resonant frequency: the note sounded when the string is plucked and the natural rhythm of a swing. The electrons bound in a crystal lattice are subject to forces analogous to the tension on a guitar string or the gravity that pulls on the swing, and so the electrons too have a resonant frequency. They can be set in motion by light, because light, for all its particle-like qualities, is also a wave that includes an oscillating electric field with a specific wavelength and frequency. If the frequency of the light is just right, the electrons resonate.

When they do, the light is strongly absorbed. But what is the connection between such absorption, based on the frequency of a light wave, and absorption based on photon energy? Just as the wavelength of light is related to its energy, so is its frequency; violet, the most energetic color, has the highest frequency, whereas red, the least energetic color, has the lowest frequency. When the quantum picture predicts that an electron will absorb a photon of higher energy, the wave picture predicts that the electron will absorb light of higher frequency.

The concept of resonance leads to a more vivid picture of absorption. Any oscillator can soak up energy when it is driven at its resonant frequency. A child being pushed on a swing goes higher and higher when the pushes match the natural rhythm of the swing; if they come at random, the swing gains virtually no height—and no energy. But once an oscillator has been set in motion, it is prey to friction. A swinging child or a vibrating guitar string loses energy to air resistance and eventually stops moving. Likewise, the energy taken up from light when electrons resonate is eventually dissipated as heat.

The idea that the electrons vibrate in response to light also leads to a deeper understanding of transparency. The frequency of visible light is too low to excite resonances in the lattice electrons of an insulator such as diamond. Yet the electrons of a transparent crystal do vibrate briefly in response to the oscillating electric field of the light. The light wave is taken up by each electron and promptly reemitted; it is passed along to the next electron in the lattice, where the same thing happens. The light is transmitted through the crystal in a kind of relay race, losing little energy along the way.

A record of these fleeting interactions can be seen in refraction. Light is refracted if it passes into or out of a transparent medium in any direction other than head-on. Crystals such as quartz and diamond refract light, though the phenomenon is more familiar in noncrystalline insulators such as glass. Refraction is what magnifies your fingers when you hold a clear paperweight; it is what causes lenses of glass or plastic to sharpen cloudy vision. The picture of how light is transmitted through an insulating crystal offers insight into refraction in any medium.

Since a light wave must pass from electron to electron in the lattice of an insulator, light travels more slowly in such a transparent medium than it does in air or in a vacuum. The slowing depends in part on how many oscillating electrons there are in each unit of volume in the solid. In addition, light slows more drastically as its frequency approaches the resonant frequency of the electrons. Thus both the electronic structure of a crystal and the frequency of the light set its speed in a crystal. And it is the change in speed, as a light wave passes from the air into a transparent solid or back out again, that results in refraction.

Picture a row of soldiers marching obliquely into a field of mud. As the soldiers at one end of the row encounter the mud, they slow down. Meanwhile the soldiers at the other end are still marching at full speed on *terra firma*. The difference in speed causes the entire row to wheel around until all the soldiers are in the mud, marching off in a new direction. The reverse of the process reorients the soldiers at the far side of the mud. In the same way, a light wave crossing the surface of a clear solid at any angle other than perpendicular is reoriented, or refracted.

Because the change in speed depends in part on the frequency of the light, refraction varies for different colors. Violet and blue, the highest-frequency visible colors, are bent more sharply than the low-frequency colors red and yellow, causing each color in white light to emerge from the solid at a slightly different angle. The dispersion explains how a prism can refine sunlight into its constituent colors and how a gem-cut diamond can turn lamplight into colorful fire.

Metals, unlike insulators, never let light penetrate their crystal lattice; even the whisper-thin layer of aluminum at the back of a mirror shrugs off virtually all the light striking it. How then can oscillating electrons explain anything about metals? Because the conduction electrons of a metal are not bound in the crystal lattice, they cannot resonate, just as a slack guitar string cannot sound a note. Yet they still respond to the oscillating electric field of light. One might think the interplay would resemble that in an insulator: the light, unable to set the electrons into resonance, would be reradiated promptly without losing energy.

That picture is accurate as far as it goes, but of course the result in a metal is quite different: instead of passing unhindered through the lattice, the light wave is reflected. The explanation lies in a further detail about how the electrons at the surface of a metal respond to the light. When they wiggle under its influence, their brief dance is out of step with the oscillation of the light. The light they reradiate is out of phase with waves continuing to travel into the lattice, resulting in destructive interference. Reradiated light can travel in only one direction without hindrance: back out of the crystal, as a reflected wave.

The pure white gleam of a metal such as silver or platinum is a sign that its electrons evenhandedly reject light of all colors. In contrast, the reddish cast of gold or copper shows that these metals must be absorbing certain wavelengths, skewing the reflected light toward the red end of the spectrum. How can they do so, when the conduction electrons reject rather than absorb light? The answer lies deeper in the electronic structure of these metals—in the electrons populating the next band down from the conduction band. As it happens, some photons of visible light are just energetic enough to promote those electrons into the conduction band, causing them to absorb the light. In gold the absorbed wavelengths lie in the green and blue regions of the spectrum, leaving red and yellow light to be reflected from the metal and returned to the eye.

This scheme for understanding how electronic structure gives insulators and metals their trademark appearances also accounts for the varied guises of semiconductors, which include common minerals as well as the key materials in many technologies. Semiconductors with large gaps between the top, filled energy level and the next, vacant level resemble true insulators. Pure zinc oxide has a band gap of 3.2 electron volts; it can absorb only the most energetic photons in visible light and so appears transparent. Silicon, the semiconductor in computer chips, has a band gap of only 1.1 electron volts, less than the energy of even the reddest visible photons. Thus silicon can absorb every color, preventing light from penetrating far into the material. Like a metal, silicon reradiates much of the light in the only direction it can go—back out of the crystal—yielding a duller version of a metal's luster.

Between those extremes are semiconductors whose band gaps fall within the energy range of visible light. Like the chromium in ruby, the intermediate-gap semiconductors absorb specific parts of the visible spectrum, allowing other colors to reach the eye. Several of these showy semiconductors have come into use as pigments, among them are cadmium yellow (cadmium sulfide) and vermilion (mercuric sulfide). Vermilion occurs in nature as the mineral cinnabar, often displayed as brick red crystals in mineral collections.

Outside such collections, the look of the world most people inhabit is the result of artifice. By applying pigments, by alloying and polishing metals, and by synthesizing new materials, human beings have been changing the look of things ever since the days of cave paintings. Now a deep understanding of the behavior of light in solid matter has made it possible to exercise far subtler control over optical properties. Materials scientists now routinely tamper with the electronic structures of semiconductors. The payoff is not just new electronic materials, to be harnessed in microelectronics, but also new tools for manipulating light. The development of materials that can generate and respond to light pulses at specific wavelengths has helped bring about a technology for communicating and, eventually, perhaps even computing with light.

The red LED that is probably winking at you from somewhere in the room; the optical transmitters, repeaters, and detectors that send telephone conversations flickering down fiber-optic cables in pulses of light; and the minute laser that reads the digital code on a compact disk are all fruits of this new technology called photonics. What makes such devices possible is the ability to control the gap between the uppermost filled energy band in a semiconductor and the conduction band above it. The band gap not only is the key to light absorption, it also governs the wavelength of light emitted when electrons in special semiconductor devices cascade across the gap, from the conduction band to the next-lower band. (Such emission is just absorption in reverse.)

The width of the band gap can be altered in many ways. A dopant—an impurity added to a semiconductor at concentrations of a few parts per million—can create new energy levels within the gap, just as do the impurities that color gems. Dopants lower the energy barrier to conduction; they also enable the semiconductor to absorb and emit light at new frequencies. Altering the basic crystal lattice is another way of controlling the band gap. By creating semiconductor compounds not found in nature or by alloying one compound with another, workers in effect devise new lattices, with distinctive band structures. In one standard material for LEDs, an alloy of the semiconductors gallium arsenide and gallium phosphide, the lattice spacing changes steadily as the proportion of the second compound is increased. At the same time, the band gap widens gradually from 1.4 to 2.2 electron volts. A gallium phosphide content of 40 percent, which sets the band gap at about 1.9 electron volts, is responsible for the red glow of many LEDs.

Sophisticated new techniques for manipulating materials have also made it possible to specify a band gap by assembling a lattice directly, atom by atom. In so-called superlattices, different semiconducting compounds are laid down in alternating layers only a few atoms thick. The scale of the layers is so close to the spacing of the atoms that the layers themselves can interact with the electron waves, influencing which ones survive and which are canceled out. As a result, the band gaps of a superlattice can be tailored just by altering the thickness of the layers.

The new materials attest to the power that comes with a deep understanding of the geometry of materials and how it shapes their appearance. But that understanding also holds spiritual rewards, foreshadowed for me in childhood by the gem collection at the natural history museum. Crystal shapes, and the periodic lattices that underlie them, seem to tap our basic need for order. They reassure us that nature need not be chaotic and has its own lucid, controlled patterns. One can imagine the delights of abandoning the human scale and entering a crystal, to live within the harmonious regularity of the lattice. One could pass an eternity in contemplation of the atomic galleries and arcades, receding in all directions. True, some see such microscopic perfection as inhuman and remote. Crystal structure, George Bernard Shaw has written, "is dangerously fascinating to thinkers oppressed by the bloody disorders of the living world. Craving for purer subjects of thought, they find [happiness] in the contemplation of crystals. . .because they see in the crystals beauty and movement without the corrupting appetites of fleshly vitality."

Perhaps one should resist the allure of crystals, to avoid losing sight of the world's rich disorder. For all the insight they offer into the optical properties of other materials, perfect atomic lattices would seem, after all, to be the hallmark of a limited set of solids, making up a small part of the universe. But a recent mathematical insight into general relativity, Einstein's theory of gravity, suggests that at the deepest level reality itself may have the texture of a crystal. It is said that the fabric of spacetime, the tapestry on which every object and event in the universe is recorded, may be made up of linked quantum loops set in a regular, three-dimensional lattice. When we contemplate the intersection of structure and light in crystals, both our true eyes and our mind's eye may be seeing in microcosm that geometry is all.

Light Dawns

If you visit the Paris Observatory on the left bank of the Seine, you'll see a plaque on its wall announcing that the speed of light was first measured there in 1676. The odd thing is, this result came about unintentionally. Ole Roemer, a Dane who was working as an assistant to the Italian astronomer Giovanni Domenico Cassini, was trying to account for certain discrepancies in eclipses of one of the moons of Jupiter. Roemer and Cassini discussed the possibility that light has a finite speed (it had typically been thought to move instantaneously). Eventually, following some rough calculations, Roemer concluded that light rays must take 10 or 11 minutes to cross a distance "equal to the half-diameter of the terrestrial orbit."

Cassini himself had had second thoughts about the whole idea. He argued that if finite speed was the problem, and light really did take time to get around, the same delay ought to be visible in measurements of Jupiter's other moons—and it wasn't. The ensuing controversy came to an end only in 1728, when the English astronomer James Bradley found an alternative way to take the measurement. And as many subsequent experiments have confirmed, the estimate that came out of Roemer's original observations was about 25 per cent off. We have now fixed the speed of light in a vacuum at exactly 299,792.458 kilometers per second.

Why this particular speed and not something else? Or, to put it another way, where does the speed of light come from?

Electromagnetic theory gave a first crucial insight 150 years ago. The Scottish physicist James Clerk Maxwell showed that when electric and magnetic fields change in time, they interact to produce a travelling electromagnetic wave. Maxwell calculated the speed of the wave from his equations and found it to be exactly the known speed of light. This strongly suggested that light was an electromagnetic wave—as was soon definitively confirmed.

A further breakthrough came in 1905, when Albert Einstein showed that c, the speed of light through a vacuum, is the universal speed limit. According to his special theory of relativity, nothing can move faster. So, thanks to Maxwell and Einstein, we know that the speed of light is connected with a number of other (on the face of it, quite distinct) phenomena in surprising ways.

But neither theory fully explains what determines that speed. What might? According to new research, the secret of c can be found in the nature of empty space.

Until quantum theory came along, electromagnetism was the complete theory of light. It remains tremendously important and useful, but it raises a question. To calculate the speed of light in a vacuum, Maxwell used empirically measured values for two constants that define the electric and magnetic properties of empty space. Call them, respectively, ε_0 and μ_0.

The thing is, in a vacuum, it's not clear that these numbers should mean anything. After all, electricity and magnetism actually arise from the behavior of charged elementary particles such as electrons. But if we're talking about empty space, there shouldn't be any particles in there, should there?

This is where quantum physics enters. In the advanced version called quantum field theory, a vacuum is never really empty. It is the "vacuum state," the lowest energy of a quantum system. It is an arena in which quantum fluctuations produce evanescent energies and elementary particles.

What's a quantum fluctuation? Heisenberg's uncertainty principle states that there is always some indefiniteness associated with physical measurements. According to classical physics, we can know exactly the position and momentum of, for example, a billiard ball at rest. But this is precisely what the uncertainty principle denies. According to Heisenberg, we can't accurately know both at the same time. It's as if the ball quivered or jittered slightly relative to the fixed values we think it has. These fluctuations are too small to make much difference at the human scale; but in a quantum vacuum, they produce tiny bursts of energy or (equivalently) matter, in the form of elementary particles that rapidly pop in and out of existence.

These short-lived phenomena might seem to be a ghostly form of reality. But they do have measurable effects, including electromagnetic ones. That's because these fleeting excitations of the quantum vacuum appear as pairs of particles and antiparticles with equal and opposite electric charge, such as electrons and positrons. An electric field applied to the vacuum distorts these pairs to produce an electric response, and a magnetic field affects them to create a magnetic response. This behavior gives us a way to calculate, not just measure, the electromagnetic properties of the quantum vacuum and, from them, to derive the value of c.

In 2010, the physicist Gerd Leuchs and colleagues at the Max Planck Institute for the Science of Light in Germany did just that. They used virtual pairs in the quantum vacuum to calculate the electric constant ε_0. Their greatly simplified approach yielded a value within a factor of 10 of the correct value used by Maxwell—an encouraging sign! This inspired Marcel Urban and colleagues at the University of Paris-Sud to calculate c from the electromagnetic properties of the quantum vacuum. In 2013, they reported that their approach gave the correct numerical value.

This result is satisfying. But it is not definitive. For one thing, Urban and colleagues had to make some unsupported assumptions. It will take a full analysis and some experiments to prove that c can really be derived from the quantum vacuum. Nevertheless, Leuchs tells me that he continues to be fascinated by the connection between classical electromagnetism and quantum fluctuations, and is working on a rigorous analysis under full quantum field theory. At the same time, Urban and colleagues suggest new experiments to test the connection. So it is reasonable to hope that c will at last be grounded in a more fundamental theory. And then—mystery solved?

Well, that depends on your point of view.

The speed of light is, of course, just one of several "fundamental" or "universal" physical constants. These are believed to apply to the entire universe and to remain fixed over time. The gravitational constant G, for example, defines the strength of gravity throughout the universe. At small scales, Planck's constant h sets the size of quantum effects and the tiny charge on the electron e is the basic unit of electricity.

The numerical values of these and other constants are known to excruciating precision. For instance, h is measured as $6.626070040 \times 10^{-34}$ joule-second (to within 10^{-6} per cent!). But all these quantities raise a host of unsettling questions. Are they truly constant? In what way are they "fundamental?" Why do they have those particular values? What do they really tell us about the physical reality around us?

Whether the "constants" are really constant throughout the universe is an ancient philosophical controversy. Aristotle believed that the Earth was differently constituted from the heavens. Copernicus held that our local piece of the universe is just like any other part of it. Today, science follows the modern Copernican view, assuming that the laws of physics are the same everywhere in spacetime. But an assumption is all this is. It needs to be tested, especially for G and c, to make sure we are not misinterpreting what we observe in the distant universe.

It was the Nobel Laureate Paul Dirac who raised the possibility that G might vary over time. In 1937, cosmological considerations led him to suggest that it decreases by about one part in 10 billion per year. Was he right? Probably not. Observations of astronomical bodies under gravity do not show this decrease, and so far there is no sign that G varies in space. Its measured value accurately describes planetary orbits and spacecraft trajectories throughout the solar system, and distant cosmic events, too. Radio astronomers recently confirmed that G as we know it correctly describes the behavior of a pulsar (the rapidly rotating remnant of a supernova) 3,750 light years away. Similarly, there seems to be no credible evidence that c varies in space or time.

So, let's assume that these constants really are constant. Are they fundamental? Are some more fundamental than others? What do we even mean by "fundamental" in this context? One way to approach the issue would be to ask, "What is the smallest set of constants from which the others can be derived?" Sets of two to 10 constants have been proposed, but one useful choice has been just three: h, c, and G, collectively representing relativity and quantum theory.

Only the dimensionless constants are really "fundamental," because they are independent of any system of measurement.

In 1899, Max Planck, who founded quantum physics, examined the relations among h, c, and G and the three basic aspects or dimensions of physical reality: space, time, and mass. Every measured physical quantity is defined by its numerical value and its dimensions. We don't quote c simply as 300,000, but as 300,000 kilometers per second, or 186,000 miles per second, or 0.984 feet per nanosecond. The numbers and units are vastly different, but the dimensions are the same: length divided by time. In the same way, G and h have, respectively, dimensions of [length^3/(mass × time^2)] and [mass × (length^2/time)]. From these relations, Planck derived "natural" units, combinations of h, c, and G that yield a Planck length, mass, and time of 1.6×10^{-35} meters, 2.2×10^{-8} kilograms, and 5.4×10^{-44} seconds. Among their admirable properties, these Planck units give insights into quantum gravity and the early universe.

But some constants involve no dimensions at all. These are so-called dimensionless constants—pure numbers, such as the ratio of the proton mass to the electron mass. That is simply the number 1836.2 (which is thought to be a little peculiar because we do not know why it is so large). According to the physicist Michael Duff of Imperial College London, only the dimensionless constants are really "fundamental," because they are independent of any system of measurement. Dimensional constants, on the other hand, "are merely human constructs whose number and values differ from one choice of units to the next."

Perhaps the most intriguing of the dimensionless constants is the fine-structure constant α. It was first determined in 1916, when quantum theory was combined with relativity to account for details or "fine structure" in the atomic spectrum of hydrogen. In the theory, α is the speed of the electron orbiting the hydrogen nucleus divided by c. It has the value 0.0072973525698, or almost exactly $1/137$.

Today, within quantum electrodynamics (the theory of how light and matter interact), α defines the strength of the electromagnetic force on an electron. This gives it a huge role. Along with gravity and the strong and weak nuclear forces, electromagnetism defines how the universe works. But no one has yet explained the value $1/137$, a number with no obvious antecedents or meaningful links. The Nobel Prize–winning physicist Richard Feynman wrote that α has been "a mystery ever since it was discovered. . . a magic number that comes to us with no understanding by man. You might say the 'hand of God' wrote that number, and 'we don't know how He pushed his pencil.'"

Whether it was the "hand of God" or some truly fundamental physical process that formed the constants, it is their apparent arbitrariness that drives physicists mad. Why these numbers? Couldn't they have been different?

One way to deal with this disquieting sense of contingency is to confront it head-on. This path leads us to the anthropic principle, the philosophical idea that what we observe in the universe must be compatible with the fact that we humans are here to observe it. A slightly different value for α would change the universe; for instance by making it impossible for stellar processes to produce carbon, meaning that our own carbon-based life would not exist. In short, the reason we see the values that we see is that, if they were very different, we wouldn't be around to see them. QED. Such considerations have been used to limit α to between 1/170 and 1/80, since anything outside that range would rule out our own existence.

But these arguments also leave open the possibility that there are other universes in which the constants are different. And though it might be the case that those universes are inhospitable to intelligent observers, it's still worth imagining what one would see if one were able to visit.

For example, what if c were faster? Light seems pretty quick to us, because nothing is quicker. But it still creates significant delays over long distances. Space is so vast that aeons can pass before starlight reaches us. Since our spacecraft are much slower than light, this means that we might never be able to send them to the stars. On the plus side, the time lag turns telescopes into time machines, letting us see distant galaxies as they were billions of years ago.

If c were, say, 10 times bigger, a lot of things would change. Earthly communications would improve. We'd cut the time lag for radio signals over big distances in space. NASA would gain better control over its unmanned spacecraft and planetary explorers. On the other hand, the higher speed would mess up our ability to peer back into the history of the universe.

Or imagine slow light, so sluggish that we could watch it slowly creep out of a lamp to fill a room. While it wouldn't be useful for much in everyday life, the saving grace is that our telescopes would carry us back to the Big Bang itself. (In a sense, "slow light" has been achieved in the lab. In 1999, researchers brought laser light to the speed of a bicycle, and later to a dead stop, by passing it through a cloud of ultra-cold atoms.)

These possibilities are entertaining to think about—and they might well be real in adjacent universes. But there's something very intriguing about how tightly constructed the laws of our own universe appear to be. Leuchs points out that linking c to the quantum vacuum would show, remarkably, that quantum fluctuations are "subtly embedded" in classical electromagnetism, even though electromagnetic theory preceded the discovery of the quantum realm by 35 years. The linkage would also be a shining example of how quantum effects influence the whole universe.

And if there are multiple universes, unfolding according to different laws, using different constants, anthropic reasoning might well suffice to explain why we observe the particular regularities we find in our own world. In a sense it would just be the luck of the draw. But I'm not sure this would succeed in banishing mystery from the way things are.

Presumably the different parts of the multiverse would have to connect to one another in specific ways that follow their own laws— and presumably it would in turn be possible to imagine different ways for those universes to relate. Why should the multiverse work like this, and not that? Perhaps it isn't possible for the intellect to overcome a sense of the arbitrariness of things. We are close here to the old philosophical riddle, of why there is something rather than nothing. That's a mystery into which perhaps no light can penetrate.

Nobody Knows the Quantum

Note: The films discussed here can be viewed at www.labocine.com.

Quantum mechanics has been around for over a century and has shown beyond doubt that it works, but its enigmas still frustrate physicists. In 2011, the eminent Swiss quantum experimentalist Anton Zeilinger asked 33 quantum theory specialists a set of 16 questions about its fundamental interpretation. Not one of the multiple-choice answers was chosen by all the participants and many questions elicited wildly varying opinions.

Despite the fact that even the professionals don't agree on the meaning of quantum theory, or maybe because of it, the quantum has become embedded in general usage and popular culture. The word "quantum" shows up in places that have nothing to do with physics, like the Quantum Theatre in Pittsburgh and Quantum fishing equipment. Quantum physics has been dubiously credited with superpowers such as improving health, underlying human consciousness, and giving the human mind the ability to alter reality—all feeding into a kind of new age–like approach that makes a hard subject even harder to get straight.

Still some LaboCine filmmakers have bravely tackled the challenge of making quantum theory and the submicroscopic world it describes more understandable or at least, less weird. If I were teaching a film-based quantum mechanics course, I'd start with *Abbau* (Masahiro Ohsuka, 2013), a five minute–long *tour de force*. It takes us from the classical physics of Newton's mechanics, optics, and electromagnetic theory through nuclear fission and fusion, quantum theory and string theory in a fluid, rapid-fire sequence of words, equations, symbols, drawings, and animations driven by a propulsive soundtrack. It all happens too fast to get details, but you're not supposed to. What you get instead is a breathtaking tour of the classical physics that quantum theory uprooted, and reminders of what is so puzzling about the theory.

Filmmaker Markos Kay more directly presents the submicroscopic universe itself through animation. *The Flow* (2012) packs a lot of ambition into its one-minute running time as it "looks at the supervening layers of reality. . .from quarks to nucleons to atoms and beyond. The deeper we go into the foundations of reality, the more it loses its form, eventually becoming a pure mathematical conception." Kay is right: what we know about the ultimate quantum scale is expressed mathematically. Kay supplements the math with a visual interpretation of the trajectory from quarks to protons and neutrons, atoms and finally molecules, all represented by sinuously moving rounded forms that resemble exotic sea creatures. But life began and operates at the molecular level, not the quark level. Even in creative imagination, it feels wrong to represent the ascent from quarks to molecules with biological images.

Kay's *Quantum Fluctuations* (2016) is more to the point. It is an interpretation of experiments at the Large Hadron Collider (LHC), the huge particle accelerator at CERN that smashes beams of high-energy protons into each other. The film's images, based on consultations with LHC scientists, have a fascinating, often beautiful complexity that could well arise from streams of protons colliding and rebounding to create new kinds of particles. Nevertheless, I don't think these two films will give anyone a clear image of quantum processes; but Kay provides a helpful bonus with an annotated "director's cut" of *The Flow* and further discussion of *Quantum Fluctuations* that give deeper insight into the microworld he explores.

Quantum mechanics seems odd because it works at a level far below the human scale, so in a way it is fitting that no people actually appear onscreen in *Abbau*, *The Flow*, and *Quantum Fluctuations*. But quantum theory is still a product of the human mind. Some filmmakers have explored the human approach to the quantum, as in Dave Fischer's black-and-white film *(a)symmetry* (2015). Against a background of shimmering shapes of light, and music that also seems to shimmer, we hear a quiet voice talking about quantum mechanics and its strange features. One of them, non-locality, is perhaps the strangest: it is the phenomenon that two quantum particles like electrons or photons, placed very far apart and without any known connection, can still affect each other.

Only near the end of the three-minute film do we learn that the speaker is David Bohm, a theoretical physicist famous for his own quantum interpretation. The voice-over in the film is an excerpt from a long videotaped interview Bohm gave at the Niels Bohr Institute in Copenhagen in 1989, where you can see as well as hear him carefully and thoroughly describe his ideas. These differ from mainstream quantum views but are not necessarily wrong; in fact they offer new insights, illustrating again, like Zeilinger's survey, that physicists are still a long way from understanding their own theory.

What can't be understood can't be explained, as is slyly and amusingly but tellingly shown in *Bien Heureux* (*All is Well*, Pierre-Arnaud Lime, 2016). An intense young man with a beard and a hat I'll call the Physicist works hard to explain quantum non-locality (here called quantum teleportation) to a friend at a party where a girl is singing, then on a Paris street as people walk by, then in a bar. The Physicist first tries the metaphor of Schrodinger's cat with a toy cat in a cardboard box, but the friend doesn't get it. Then the Physicist uses two boxes with colored dots to represent two separated particles, but the friend remains clueless.

Finally, in his last attempt, the Physicist explains using a laser pointer and the interference of light. The people grouped around him in the bar all nod "yes, we understand"—that is, all but the friend. The Physicist recoils in disappointment, but the final scene shows the two sitting on the bank of the Seine, amiably trading snatches of poetic word imagery and then parting, still friends but with quantum teleportation still a mystery.

These five films show a variety of artistic approaches to presenting quantum physics. We should not expect an artwork to give the same results as a physics lecture, but films like these can give something different and valuable: a sense of how distant the quantum world is from ordinary experience, and also the reassurance that you, trying to grasp the theory, are not alone because nobody, truly nobody, understands the quantum.

Strange Devices

Look around the room in which you are reading this book. You and all that surrounds you—your chair, the ink on this page, the air you breathe—are made up of atoms. And every atom—every particle out of which atoms are made—is, in turn, subject to the laws of quantum physics, the mysterious rules that describe a physical reality radically different from the expectations of everyday common sense. Here is a world in which solid matter undulates. Particles ooze out of leakproof containers. Light ejects bullets of pure electricity out of hard matter as the color of the light turns from red to blue.

Intuitive judgment fails in the quantum world, because people have almost no direct experience with quanta. To catch a hard-hit ball to center field, a ballplayer relies on the fact that the ball acts in a way physicists call classical: it obeys the laws of motion Newton set forth 300 years ago. But examine your surroundings more closely, and you will find human-made systems that display decidedly nonclassical effects. Within the circle of your sight or your reach, it is not unlikely that there is a piece of electronic equipment. Whether the equipment is a commonplace digital watch or a complicated stereo system, inside it are artificial devices made from semiconducting materials. Such devices manipulate electrons and photons, the units of electricity and light, for desired ends. In contrast with the classical simplicity of the larger, macroscopic world, the physical laws that dominate those minute devices are quantum laws, and their performance is controlled and enhanced by the manipulation of the quantum world.

Earlier generations of electronic devices, such as vacuum tubes for radios and the cathode-ray picture tubes still ubiquitous in television and computer-monitor screens, can be understood without quantum theory. But the functioning of more recently developed devices depends on quantum effects in the semiconducting materials out of which the devices are made. The chips made of silicon at the heart of your computer or in the fuel-emissions controller of your car, and the silent, minuscule laser made of gallium aluminum arsenide that sends your voice through fiber-optic telephone lines, all rely on quantum effects for their operation. And new designs that take advantage of exotic quantum phenomena could make the next generation of devices even smaller, faster, and more efficient. Such designs make sense because it is now feasible to fabricate systems so small that only the quantum effects are important. What are those new designs? Which of the many weird quantum effects are most likely to be useful? How can their application be expected to change our lives? There is much buzz in the air these days about nanotechnology—a technology that operates in the atomic dimensions characterized by the nanometer, a billionth of a meter. But a better term for the field might be quantum technology: a new discipline of design and engineering that can take full advantage of the strange world of the quantum.

The laws that govern the quantum universe can only be called mandates for ambiguity. Early inklings of that ambiguity came as the nineteenth century ended, when evidence mounted that the century-old model of light as an undulating wave could not explain its interaction with matter. The German physicist Max Planck first proposed that energy comes in steps, or quanta, rather than as a smooth flow. Building on Planck's work, Einstein proposed in 1905 that light is also a quantum of energy, a particle later given the name photon. That startling concept established the need to regard light—and indeed, all electromagnetic radiation—as sometimes a wave and sometimes a particle.

By a reciprocal logic, the duality of particle and wave was extended not long thereafter from the description of radiation to a description of matter. In 1923 the French physicist Louis-Victor de Broglie derived a deceptively simple equation, eerily implying that every bit of matter is something more than solid, localized reality. Bits of matter also undulate: they are waves. De Broglie's equation involves momentum, which is a classical property of matter (namely, the product of a particle's mass and velocity), and wavelength, an intrinsic property of waves (the distance between two adjacent wave crests). Given a particle with a certain momentum, the equation shows how to determine the particle's wavelength. In so doing, the equation forges a baffling link between particles and waves that still haunts the quantum physics of matter.

The wave aspect of matter leads to the quantum properties applied in devices. It is the reason electrons mutually interfere, like so many water waves; it is responsible for the indeterminacy of the microscopic world that violates any sense of a clockwork universe; and it is the basis for the quantization of energy in solids. Paradoxically, for all that, it is the quantum fuzziness of matter that can so dramatically improve device performance.

The most powerful evidence for de Broglie's matter waves is the interference of electrons, first observed in 1927. Interference is the merging of two or more waves at the same time and place, and it highlights a profound difference between particles and waves. The effect of multiple particles converging in time and space is always greater than the effect of one particle. Two tennis balls hitting a racket deflect it more than one ball does; two cannon balls are better than one for battering down a castle wall. But the effect of multiple waves is much more complex, because it can be either stronger or weaker than a single wave.

Imagine waves being generated in a pool of water by a single source—say, a block of wood on a spring—then separated into two distinct channels. The moving peaks and valleys in each channel start in phase—that is, the crests and troughs of each wave train move along each channel in unison. Suppose the channels finally spill back into another pool where the waves reunite. If the channels are the same length, the crests and troughs of the wave trains are still in perfect step when the waves rejoin. The net effect is that crest adds to crest and trough adds to trough, doubling the height of each wave; the outcome is called constructive interference.

But suppose one channel is longer than the other by half a wavelength. Then when the waves recombine, the crest of one just matches the trough of the other, and the two waves cancel out—an effect known as destructive interference. Particles can never destroy each other in that way, and so the phenomenon clearly distinguishes waves from particles. Only recently has it become possible to control constructive and destructive interference in ordinary matter well enough to exploit it in devices.

The wave nature of matter is also responsible for the uncertainty principle of Werner Heisenberg. The name of the principle is its message: certain physical data are simply unknowable. For example, it is not possible to find both the position and the momentum of a particle simultaneously. In fact, the more known about one, the less known about the other. The circumstance has a certain ironic power, not unlike the tale of the "appointment in Samarra": a man's strenuous efforts to flee Death bring him to the very spot where

Death finds him. Similarly, according to the uncertainty principle, the harder one tries to outwit the intrinsic elusiveness of physical reality, the more surely one will have to confront it face to face.

Everyday activities, of course, are not plagued by the uncertainty principle. A ballplayer instantaneously judges both the momentum and the position of a ball to make an elegant catch. No one— Ghostbusters and neutrino physicists excepted—is familiar with objects that pass through walls. But the only real reason such odd behavior is not more common is quantitative, not qualitative: the uncertainties defined by Heisenberg's principle are, by everyday standards, infinitesimal. Compared with the momentum and dimensions of any macroscopic object, from pea to pachyderm, the uncertainties are far too small to measure. But in the subatomic world they loom so large that they strongly affect electrons.

Perhaps the most important implication of matter waves for device physics is the quantization of energy. In the simplest atom, hydrogen, a single electron orbits a central proton. The crests and troughs of the orbiting electron's de Broglie wave can be imagined as waves generated on a circular loop of string. If such a loop were plucked and made to vibrate, the only wavelengths that could exist would be the ones that fitted exactly, an integral number of times, into the circumference of the loop. Other wavelengths would simply die out immediately after they were born, because the peaks and valleys of the waves would cancel one another.

Similarly, only a certain de Broglie wavelength is permitted for each electron orbit, depending on the size of the orbit. Since, in de Broglie's equation, wavelength determines momentum, and because momentum determines energy, each orbit is associated with a specific energy. The quantization of energy, in which each orbit is associated with a particular energy level, is the heart of the quantum-mechanical description of the hydrogen atom or, indeed, the atom of any other element.

The magnetic properties of electrons are quantized as well. Picture the electron, as one does in classical physics, as a minute sphere of charge. If the sphere spins on its axis, the moving charge constitutes an electric current. Flowing current makes magnetism, and so a spinning electron acts like a minuscule bar magnet whose north and south poles are oriented along its axis of spin. Any such magnet can be affected by any other magnet.

That classical picture was tested in 1925: a stream of electrons was passed through a magnetic field, then their impact was detected on a screen. The electrons were magnetically deflected from their paths, but in a surprising manner. Instead of a continuous range of deviations, as one would expect from a randomly oriented swarm of spinning electrons, each electron was deflected in only one of two ways: a specific distance up from its initial line of travel, or the same distance down. In other words, the electrons acted as if their spin was quantized; their spin axis seemed to point either up or down, and never at any other angle with respect to the magnetic field.

An understanding of strange quantum devices begins, however, with the classical view of electrons in a conducting medium say, a copper wire. The atoms in the wire assemble themselves into a regular, three-dimensional array, each accompanied by a family of electrons. What makes copper a good electrical conductor is that some of the electrons in each atom are free to roam within the atomic structure.

Imagine yourself compressed by a factor of 100 billion (a shrinkage that would reduce the Earth to a dot smaller than the period at the end of this sentence), and look about at the roaming electrons. Perch on a copper atom and peer in all directions at a nanoworld of static architecture combined with frenzied action. The architecture is framed by the massive copper atoms, each with a diameter of many feet relative to your imagined size. Their symmetrical arrangement, extending to the horizon, contrasts with the great random swarm of much smaller free electrons. Moving at high speed, the electrons carom off the atoms like so many billiard balls or the molecules of a gas. Classical theory explains what happens to such an electron gas if the wire that contains it is connected across a battery. The negative pole of the battery repels electrons, while its positive pole attracts them, thereby adding an ordered though surprisingly slow net velocity to the random electronic motion. The result is a flow of charge, an electric current.

To extend classical understanding into the quantum world, perch again on your copper atom, this time gazing at the electronic world through eyes informed by quantum theory. Now you see a complex atomic structure filled with undulating matter waves rather than ricocheting electrons. It is as if a long-sunken city of graceful classical arcades, Atlantis, if it ever was, were buffeted by subsurface storms that agitated the water filling its drowned plazas and marketplaces. Examine the great profusion of electronic waves for a time, and you see a subtle pattern. Just as in the quantized electronic orbits of individual atoms, certain wavelengths never appear, because the regularity of the atomic lattice imposes geometric conditions on the waves. The ones that do not exactly fit the atomic architecture undergo destructive interference and soon die out.

According to de Broglie's equation, the missing electronic wavelengths correspond to missing values of momentum, which implies that certain energies never appear. The significance of the missing energies can be appreciated if one takes another look at the quantum nature of a single atom. Each of its electrons occupies a specific orbit with a definite energy; the farther the electronic orbit is from the atomic nucleus, the higher the energy of the electron. Electrons ascend and descend among the orbital energy levels as if they were rungs on a ladder. If an electron in a low-lying inner orbit gains energy, it jumps to a higher empty orbit, leaving behind an empty track.

The same quantum idea of high- and low-energy levels, separated by a forbidden energy gap, applies to a semiconductor. But whereas each energy rung in the atom accommodates only two electrons, the energy levels in the semiconductor include all the vast number of electrons that reside in a solid of macroscopic size. The electrons below the gap occupy states of relatively low energy collectively known as the valence band. Those electrons are tightly bound to the atoms in the semiconductor and cannot contribute to the flow of electric current. Electrons above the gap reside in high-energy states called the conduction band, and those electrons are relatively mobile. Just as electrons in an atom jump from rung to rung, energetic electrons in a semiconductor can leap from valence to conduction band.

The structure of band and gap defines the use of semiconductors in devices. An electron promoted from valence band to conduction band leaves behind a space, unpoetically but aptly called a hole. The hole makes room in which the remaining valence electrons can move—much like the empty space in the common children's puzzle in which fifteen tiles are to be shifted around within a square frame that has room for exactly sixteen tiles. The space left by the omitted tile is what makes motion possible, and the hole in the valence band enables the electrons that remain in that band to join in the flow of current. Since missing negative charge is the same as added positive charge, the holes act like mobile positive charges.

Semiconductors such as pure silicon can be "doped" with impurities chosen to add either extra electrons in the conduction band or holes in the valence band. When the electrons dominate, the material is called an *n*-type (for negative) semiconductor; when the holes dominate, the material is *p*-type (for positive). When the two types are butted together to make a *p-n* junction, the result is a useful device called a diode, a one-way street that resists the flow of electrons in one direction but not the other. That property can, for instance, change alternating current, the kind that comes from wall outlets, into unidirectional direct current.

A diode has two elements. More important still are arrangements of semiconductors in which the number of elements is three. The combination of three adjacent regions in the sequence *n-p-n*—like the wafers and the filling of an Oreo cookie—is one of the earliest forms of transistor, whose invention, in 1948, launched the electronic age. Its value is that a small voltage applied to the *p*-type region controls a large flow of electrons between the two *n*-type regions; in the terminology of electronics, a transistor is a three-terminal device in which the voltage applied to the gate modulates a large current between source and drain. Hence electronic information coming in to the gate at low power, such as the complex waveform of an audio or video signal, is amplified into a stronger but otherwise identical pattern in time. Transistors have thus become the basis of stereo and television receivers.

Almost any piece of electronic equipment is now so complex that it requires thousands or millions of such basic amplifiers and switches. Those parts are made as small as possible and built into a single small chip of semiconductor, usually silicon. But the technology is reaching its limits. The largest commercially available computer memory chip now holds 16 million bits (megabits) of data. To reach a billion bits (one gigabit) per chip, devices must become much smaller. But with such dense packing, even a minute amount of power supplied to each device can generate more heat than the equipment can tolerate. Another factor limiting the performance of electronic equipment is the speed of electrons within a device, or from device to device. The direct application of quantum principles offers the designer a variety of ways to enhance those measures of performance.

One step needed to manufacture quantum devices has already been taken: it is possible to make exceedingly small structures out of semiconductors and metals. New techniques for manipulating atoms, such as molecular-beam epitaxy and scanning-tunneling microscopy, now come close to the ideal of selecting one atom at a time and setting it in place. It has become routine to make structures with layers of known composition a few nanometers thick or with narrow conductive conduits, known as quantum wires, a few tens of nanometers across. Quantum wires are the electronic analogues of the long, narrow water channels envisioned in my earlier thought experiment about interfering waves. An electron in a quantum wire moves freely along its length, but the electron cannot slide off to the side. Experiments in which two quantum wires are merged clearly demonstrate the effects of interference, again showing electrons to be traveling matter waves, not miniature bowling balls.

Even such a simple two-channel configuration gives one a subtle control over the electrons in devices. For example, if a small electric field were to slightly affect the electrons in one channel, the two merged waves could be instantly shifted from constructive to destructive interference. In other words, the current flow would rapidly turn from "on" to "off." And there are now three-terminal quantum interference devices that function as conventional transistors. One design exploits the Aharonov–Bohm effect, discovered in 1959, in which a magnetic field affects the phase of the electron wave. In that device, the source is connected to the drain via two channels along which electron waves propagate, and a magnetic field is set up between the two channels. A change in the voltage at the gate alters the magnetic field, thereby changing the phase relation of the two waves. The resultant change in the interference pattern of the two merged waves controls the current flowing out of the drain.

Other quantum schemes enhance electron waves the way an organ pipe enhances sound waves. Just as an organ pipe of a certain length amplifies sound from a vibrating column of air, the supporting system of the electron waves can enhance those wavelengths that match the dimensions of the system. The resonant tunneling transistor treats electron waves in that dramatic fashion.

Two parallel "mirrors" made of one kind of semiconductor are set a precise distance apart; the electron mirrors act as source and drain for the transistor. Between them lies the gate, a semiconducting channel with characteristics different from those of the mirrors. Electron waves tunnel back and forth through the channel, reflected many times at each end by the semiconductor mirrors. But only the waves whose wavelengths match the distance between the mirrors (the exact condition is that an integral number of half-wavelengths fit into the cavity) can freely travel along the channel. Other electron wavelengths, which do not exactly fit into the cavity, interfere destructively with one another; thus no current passes from source to drain. Voltage applied to the channel controls the energy of the electrons in the channel and, therefore, by de Broglie's equation, their wavelengths. Thus gate voltage modulates current flow, the basic transistor effect.

The resonant tunneling transistor functions at high speed, and it has another advantage. It takes only a small change in energy to change the electronic waves in the channel from resonating to nonresonating, and so the device can operate at extremely low voltages. A reduction in operating voltage is a significant step in the further miniaturization of electronic equipment.

Mark Brian Johnson, a physicist at the Naval Research Laboratory in Washington, D. C., has proposed and tested a different kind of quantum device, based on electron spin. Johnson's invention might be called an *f-m-f* device, for it sandwiches a metal film (gold, in the prototype units) a few nanometers thick between films of ferromagnetic material, which can be strongly magnetized in specific directions. Voltage applied between one of the *f*-films and the *m*-film makes a current whose electrons all have the same spin, say, up. The magnetic condition of the other *f*-film, which can be easily changed, influences how much current finally emerges.

The device is conceptually similar to an *n-p-n* transistor. It can be used as a switching transistor or as an amplifying one, and it has one important new feature. Because the device operates magnetically, it retains a record of the last spin state to pass through it, even after power has been switched off. That property contrasts sharply with the way binary digits are represented in present-day computer memories, by the flow of electronic current or its absence. When power is removed from those memories, the memories are erased.

Johnson's device could lend itself to extreme miniaturization, because the magnetic effects become stronger as the films become thinner. And the use of metals rather than semiconductors is attractive for commercial processing.

Semiconductor devices not only manipulate electrons. They also convert electrons and photons into each other, thereby supporting the technology of light that has come to be known as photonics. Photonic devices exploit the capacity of semiconductors to emit or absorb light—a quantum phenomenon par excellence. An LED is a *p-n* junction arranged so that the flow of current brings a steady supply of electrons to the conduction band, like a river endlessly carrying water to the rim of a high cliff. Like the waterfall that overflows that rim, electrons continuously cascade down through the gap to the valence band. The electronic waterfall generates a kind of quantum spray, for just as solid water changes into ethereal fine droplets on the way down, each descending electron gives up its excess energy by emitting a photon in a "poof" of transmutation. By choosing among the band gaps of various semiconductors (sometimes modified by the addition of impurities), the light can be made in most of the colors of the spectrum. Triads of red-, green-, and blue-emitting diodes incorporated into a flat-panel screen can make a full-color display, which may soon replace the bulky picture tubes that are almost the last vestige of old-fashioned electronics.

In spite of their tremendous promise, the commercial development of quantum devices faces some serious barriers. Most fundamental is that the devices need low temperatures, because the random thermal energy of atoms at room temperature is great enough to disturb the quantum states of electrons in motion. Interference experiments with quantum wires, for instance, require temperatures too low to be feasible for general use.

Part of the answer is to seek different physical principles. The magnetic properties of electrons are less sensitive to thermal effects than are their quantum-interference properties, and spin-based devices are expected to function well at room temperature. Another possible solution takes advantage of quantum indeterminacy. As the width of a quantum wire is reduced, the position of the electron is more narrowly constrained, and so, according to the uncertainty principle, its momentum, and its energy, can be quite large. In sufficiently small devices, the electronic energies become large enough to override thermal effects. That solution, however, awaits the breakthrough in technology needed to form quantum wires and other devices perhaps fifty times smaller than it is possible to form today.

It is ironic that, despite their nearly hundred-year history, and despite their application in devices many of us virtually take for granted, quantum phenomena and their implications remain elusive. Even scientists with the keenest insight into the physical universe, the Einsteins and the Feynmans of this world, have confessed puzzlement at the paradoxes that seem to seep into every corner of the quantum world. Those who would build quantum devices today must forge ahead despite a lack of profound understanding about that world. Yet by building such devices, there is the hope and the promise of bootstrapping the basic understanding. Along with their impact on technology, the new devices can serve as miniature laboratories in which one might learn much more about the philosophical riddles that permeate the quantum universe.

To experience some of the sense of mystery that quantum phenomena can engage, I found, in seeking inspiration for this article, that there is nothing quite like experiencing the phenomena firsthand. For a small sum you can buy an assortment of millimeter-size light-emitting diodes and a battery. Connect the battery to an LED and look at the bright dot of light that glows forth. From my own collection of LEDs I could choose crisp yellow, ruby red, vibrant green, and soothing red-rose-orange. Now take note: in this small device, costing mere pennies, you are contemplating nothing less than the enigmatic process whereby the particle-waves of electrons become the wave-particles of photons. Your gaze, your thought, will probably not resolve the unanswered questions, but still there is something to be learned: that the microscopic quantum universe is full of strange and wondrous things, waiting to be explained.

These Georgia Tech Physicists Helped Prove Einstein Right

Deirdre Shoemaker has known from the time she was a 12-year-old science-fiction fan that she wanted to spend her life studying black holes. But when she came to Georgia Tech in 2008 as a founding faculty member of the university's Center for Relativistic Astrophysics, she found few other female postgraduates.

"You see women in biology, life sciences, and even math, but physics is still lagging for whatever reason," says the bubbly Shoemaker, who in 2013 became director of the center, which researches cosmic mysteries like dark matter and particle physics.

This past February, Shoemaker and Laura Cadonati, a veteran researcher who joined Tech last year, were part of the international team that confirmed the existence of gravitational waves, a long-elusive cosmic feature first predicted a century ago by Albert Einstein's theory of relativity.

A few days after the scientific breakthrough made headlines around the world, the two women delivered a presentation on the findings during a Sunday afternoon event sponsored by the Atlanta Science Tavern. Before an improbably standing-room-only crowd at the Decatur Recreation Center, they explained that the first gravitational waves ever to be detected had come from the collision and merger of two black holes—each about 30 times the mass of the sun—that occurred 1.3 billion years ago.

"The gravitational wave discovery," Cadonati says, "has opened up new ways to study the universe" because the waves can be used to collect data about distant objects like neutron stars and cosmic events like the Big Bang.

Both Shoemaker and Cadonati are part of the Laser Interferometer Gravitational-Wave Observatory (LIGO), a global collaboration of more than 1,000 scientists working to detect the waves Einstein predicted in 1916. Students of general relativity—and fans of the movie *Interstellar*—know that gravity is produced when a mass distorts the universe's underlying spacetime, which is why time slows near a source of gravity.

Gravitational waves are essentially ripples in spacetime radiating at the speed of light and can be observed as they sweep by and momentarily change the distances between objects. But the effect is too small to be detected unless the waves are especially strong, such as those emanating from a cataclysmic event like a black hole mashup, and even then the measuring tools must be extraordinarily precise.

On September 14, 2015, LIGO scientists working at identical four-kilometer-long laser detectors—located 2,400 miles apart in Washington State and Louisiana—recorded the telltale signature of a gravitational wave washing over Earth from deep space. The signal appeared seven milliseconds later in Washington, telling observers that it had come from somewhere in the southern sky. After quietly spending months corroborating the findings, the LIGO team made its historic announcement on February 11, 2016.

As chair of LIGO's Data Analysis Council, Cadonati oversees the verification of all experimental results, including the September 14 signal. Shoemaker's team at Tech crafted hundreds of binary black hole simulations. Their calculations helped establish that the detected wave had to have come from that long-ago collision.

Cadonati has found that an important part of her role in LIGO involves brokering agreement among smart, highly competitive researchers. She now jokes that her experience has prepared her for a career in politics.

A native of Italy, Cadonati, 45, earned her PhD at Princeton and taught at University of Massachusetts Amherst. She became hooked on experimentation during a year at Italy's Gran Sasso National Laboratory, where she studied the properties of neutrinos, nearly imperceptible particles emitted by the sun.

Even though Cadonati holds a high post in LIGO, men fill nearly all of the other leadership positions. She believes the scarcity of women physicists in the tops ranks of academia is due to female students' lack of self-confidence and perhaps subtle discouragement along the way. But the numbers are ticking upward: About one in five physics doctorates are currently held by women, up from one in 10 in 1990.

"I don't see hurdles when I look back at my career, and I've enjoyed support from colleagues," she says. "But I've also noticed an intangible bias against women in physics."

Her daughter, she says proudly, is in a high school honors program for math.

Shoemaker, 44, who earned her doctorate at the University of Texas in Austin, likewise says she always felt people wanted her to succeed in her chosen field of theoretical astrophysics, but concedes that female scientists often aren't taken as seriously as their young male counterparts. "I'm goofy and friendly, so that may have worked against me, but I was never discouraged," she says. "Most people thought [my interest in science] was cute and that I would grow out of it."

Three times a week, she takes part in a LIGO-related teleconference. About half her research time is devoted to the project, which reported the second detection of a gravitational wave in mid-June. The LIGO research, which is funded by the National Science Foundation, is considered a shoo-in for a Nobel Prize—at least for the group's founders at Caltech and MIT. Still, Shoemaker and Cadonati believe that involvement with LIGO will benefit and inspire their female students and junior colleagues as well.

"It's pretty astounding that the data that LIGO has collected has so closely aligned with Einstein's theory," Shoemaker says, pointing out that the supercomputers that performed the calculations proving his predictions didn't exist during his lifetime. As a bonus, gravitational waves have given scientists additional observational evidence for the existence of Shoemaker's beloved black holes, another Einsteinian prediction.

As faculty advisor to Tech's Women in Physics undergraduate social group, Shoemaker has been encouraged to see that about a third of the department's incoming grad students are women. Recently promoted to full professor, she's now one of the few senior women in academic physics. "It's lonely up here," she says.

Quantum Gravity

Watching a rocket as it slowly starts to heave itself out of Earth's deep gravity well and then streaks up into the blue, you suddenly grasp on a visceral level the energies involved in space exploration. One minute that huge cylinder is sitting quietly on its launching pad; the next, its engines fire up with a brilliant burst of light. Clouds of exhaust fill the sky, and the waves of body-shaking thunder never seem to end.

To get anywhere in space, you have to travel astounding distances. Even the Moon is about 400,000 kilometers away. And yet the hardest part—energy-wise, anyway—is just getting off the ground. Clear that hurdle, slip the bonds of Earth, and you're off. Gravity's influence falls away and suddenly, travel becomes a lot cheaper.

So it might be surprising to hear that the most exciting new frontier in space exploration starts a mere 2,000 kilometers above the terrestrial surface. We aren't talking about manned missions, automatic rovers, or even probes. We're talking about satellites. Even more prosaically, we're talking about communications satellites, in low Earth orbit. Yes, they'll be fitted with precision laser equipment that sends and receives particles of light—photons—in their fundamental quantum states. But the missions will be an essentially commercial proposition, paid for, in all probability, by banks eager to protect themselves against fraud.

Perhaps that doesn't sound very romantic. So consider this: those satellites could change the way we see our universe as much as any space mission to date. For the first time, we will be able to test quantum physics in space. We'll get our best chance yet to see how it meshes with that other great physical theory, relativity. And at this point, we have very little idea what happens next.

Let's back up. Since its discovery in 1900 and its formalization in the 1920s, quantum mechanics has remained unchallenged as our basic theory of the submicroscopic world. Everything we know about energy and matter can (in principle) be derived from its equations. In an extended form, known as quantum field theory, it underlies the "Standard Model"—which is to say, all that we know about the elementary particles.

It's difficult to overstate the explanatory power of the Standard Model. Physics has identified four fundamental forces at work in the Universe. The Standard Model accounts for three of them. It explains the electromagnetic force that holds atoms and molecules together; the strong force that binds quarks into protons and neutrons and clamps them together in atomic nuclei; and the weak force that releases electrons or positrons from a nucleus in the form of beta decay. The only thing the model leaves out is gravity, the weakest of the four. Gravity has a theory of its own—general relativity, which Albert Einstein published in 1916.

Many physicists believe we should be able to capture all our fundamental forces with a single theory. It's fair to say that this has yet to be achieved. The problem is, quantum theory and relativity are based on utterly different premises. In the Standard Model, forces arise from the interchange of elementary particles. Electromagnetism is caused by the emission and absorption of photons. Other particles cause the strong and weak forces. In a way, the micro-scale world functions like a crowd of kids pelting each other with snowballs.

Gravity is different. In fact, according to general relativity, it's not really a force at all.

Picture an empty canvas hammock stretched out flat between two trees. In places where there are no big masses (stars, for example), spacetime is a bit like that. An apple placed on the hammock would stay put. Given a shove, it would roll across the canvas in a straight line (in the real universe, that's how bodies behave when they are far away from any large masses). But when you settle *your* mass into the hammock, you distort its flatness into a dip. Given a shove, the apple would swerve around the dip—or maybe just roll straight into your side. In much the same way, when a planet orbits a sun, there's no force *pulling* it—it is simply following a curved path in distorted spacetime. Gravity is what we call that curvature. As the American physicist John Wheeler put it in 1990: "Mass tells spacetime how to curve, and spacetime tells mass how to move."

This geometrical character sets general relativity apart from quantum mechanics. They might have been lumped together as "modern physics" in the 20th century, but really, these theories merely coexist. As the British physicist J. J. Thomson wrote in 1925 in a different context, they are like a "tiger and shark, each is supreme in his own element but helpless in that of the other." And yet the prospect of bringing them together remains irresistible. It might be true that we have working theories for all the fundamental forces, but until we can unite them within a single theory, important parts of the cosmic environment and its history remain obscure.

Such as? Well, using our existing theories we can trace the history of the universe back almost to the Big Bang itself, 13.8 billion years ago. Almost but not quite: we know next to nothing about the first 10^{-43} seconds. What we can say is that the cosmos must have been extremely small and hot at that point.

That earliest "Planck epoch" is defined using tiny fundamental units invented by Max Planck—a length of about 10^{-35} meters, a mass of about 10^{-8} kg, and a duration of about 10^{-43} seconds. The Planck length, far smaller than any elementary particle or distance that we could measure, is the ultimate quantum uncertainty in location. It is the scale at which gravitational and quantum effects are equally strong. What that means is that, when the universe was still that small, it could be understood only through a theory that includes both gravity and quantum mechanics.

Such a theory is also necessary for black holes, those cosmic zones where mass is so concentrated that not even light can escape its gravitational effects. Black holes are common—they exist at the center of our galaxy and many others—and though general relativity predicts them, it does not fully describe them. In particular, it is silent about what happens at their centers, where spacetime becomes infinitely curved. So, a decent theory of quantum gravity would shine a light into some mysterious places.

Sadly, our best efforts to develop this theory remain unconvincing. One idea is that gravity is carried by hypothetical particles, just like the other three forces. These "gravitons" are predicted by string theory, in which elementary particles are quantum states of tiny vibrating strings. But string theory itself is controversial, mainly because, despite the beauty of its mathematics, it fails to generate anything resembling a testable prediction (as Einstein said: "If you are out to describe the truth, leave elegance to the tailor"). A competing theory called loop quantum gravity, in which spacetime itself has a quantum nature, also lacks testable implications.

What to do, then? In physics, mathematical beauty and ingenuity are never enough by themselves. What is needed is *data*—especially new data. And it happens that the prospects for that are suddenly very good.

Recall that quantum mechanics describes the micro-world of fundamental particles. General relativity, meanwhile, describes how celestial bodies operate over great distances—it governs the vast expanses of the cosmos. What we don't yet know is what happens to quantum phenomena over long distances. In short, we haven't tried to do quantum experiments where relativity gets in the way. Yet.

Today, the opportunity to do just that looks likely to emerge from a plan to improve our current telecommunications infrastructure. At the moment, a lot of data—internet, TV, and suchlike—is transmitted as pulses of light through a global fiber-optic network. Those pulses are made of photons, of course, but the network does not make use of the photon's exotic quantum properties. The new idea is to see what happens when it does.

Let's look at photons for a moment. Each one has an electric field that can be polarized—that is, it can be made to point in either of two directions at right angles to one another, which at the Earth's surface are "horizontal" and "vertical." We could indicate the former with a "0" and the latter with a "1." And once we're thinking about our photon that way, it's only a short leap to seeing it as a binary bit, like a switch in a computer processor. But there's a twist. A regular computer bit is always 0 *or* 1, with no other options. The quantum nature of a photon, by contrast, allows it to represent 0 *and* 1 simultaneously. It's a sort of super-bit. We call it a "quantum bit," or qubit for short.

Qubit-based computers, now being designed, are expected to far outperform ordinary computers for certain problems. And there's one area where the use of qubits is anticipated with particular eagerness: data security. In any communications system, sensitive information such as financial data can be encoded and sent to a recipient who has the key to the code. The trouble is, it's always possible for a third party to sneak into the network and secretly learn the key. It was this kind of breach, for example, that recently leaked the credit card numbers of millions of customers of U. S. retail chains such as Target and Home Depot. Qubits should prevent that.

Using a procedure called "quantum key distribution" (QKD for short), the innate quantum uncertainty about the polarization of each photon allows us to generate a long random string of 0s and

1s, which can then be sent as a totally secure key. It's secure because any interception would be detected—reading the bits changes them, thanks to the Heisenberg's uncertainty principle.

Well, that's the theory. In practice, it turns out that photon qubits cannot reliably be sent through long stretches of optical fiber. So researchers are developing an audacious new plan for a secure global data network: they intend to transmit qubits between ground stations and space satellites in low Earth orbit (LEO) at altitudes up to 2,000 kilometers.

There are groups working on this project at the Institute for Quantum Computing (IQC), part of the University of Waterloo in Canada, and at the Centre for Quantum Technologies in Singapore. The most advanced effort, supported by the Chinese Academy of Sciences, is based at the University of Science and Technology of China in Hefei (USTC). In 2017, researchers at this university launched an LOE Quantum Science satellite, equipped to test both secure quantum communications and fundamental quantum effects.

Once qubits can be exchanged with this or any other satellite, we can begin examining quantum mechanics in space. Surely the first thing to look into would be the exotic effect of *entanglement*   what Einstein called "spooky action at a distance." Entanglement means that, once two quantum particles have interacted, they remain linked no matter how far apart, so that measuring one instantly affects the other. We can, for instance, prepare a photon pair so that they are oppositely polarized (horizontal and vertical, 0 and 1), without knowing which is which. As soon as the polarization of photon 1 is measured, no matter what the result, measurement of photon 2 will reveal the other value with no physical connection between the two.

The strangeness of quantum entanglement was first laid out in the 1935 "EPR" paper by Einstein, Boris Podolsky, and Nathan Rosen. EPR argued that entanglement contradicts "local realism," which is the idea that objects have innate properties independent of measurement and that they cannot affect each other any sooner than light can travel between them. In an amazing *tour de force* in 1964, the Northern Irish theorist John Bell showed that, if data from two entangled particles obey a mathematical relationship called Bell's inequality (as it appears they do), they must violate local realism. And they must violate it "instantaneously," through a quantum feature such that "the setting of one measuring device can influence the reading of another instrument *however remote*" (emphasis mine).

However remote? Entanglement has been tested up to a distance of 143 kilometers, by Anton Zeilinger at the University of Vienna. In 2012, Zeilinger's former student Jian-Wei Pan, who heads the USTC quantum satellite effort, achieved entanglement over distances nearly as large. And then, in an important "proof of principle" demonstration in 2013, he managed to transmit photon qubits 800 kilometers from a ground station to an orbiting German satellite and back. Results like these start to make quantum research in space look very doable.

So what can we hope to investigate? Conveniently, the Institute for Quantum Computing recently published a survey of the most exciting avenues. The first experiments to push into new regimes would use LEO satellites to test entanglement up to a distance of 2,000 kilometers. If entanglement and Bell's inequality still hold, that would give the strongest evidence yet against local realism. An important result. But if either failed within that distance, then the meaning of entanglement would have to be utterly rethought. That would be even more interesting.

LEO measurements could also support a test not possible on Earth. When an observer measures one of an entangled pair of photons, in principle, it simultaneously determines the state of its partner as measured by a second observer. But in special relativity, the meaning of the word "simultaneous" is complicated. Observers moving relative to one another, who measure two physically separated events, will disagree over who made the measurement first. The time difference is too small to measure at earthly speeds, but LEO satellites move fast enough that the apparent paradox could be examined to explore the supposedly instantaneous nature of entanglement.

However, current quantum technology might be sufficient to transfer qubits and test entanglement still further out—to geostationary satellites 36,000 kilometers above the Earth's equator, and even 10 times further, to the Moon. At these distances, the curvature of spacetime starts to become important. For that reason, Thomas Jennewein, a lead author of the IQC paper, thinks experiments in this broader regime are most likely to discover something big.

Such as? One example concerns what is called gravitational redshift. We know that a photon's wavelength moves toward the red end of the spectrum as the photon climbs upward against gravity. The reason is that time dilates—that is, clocks literally run slower—in an intense gravitational field. Even in the short climb to a LEO satellite, photon qubits ought to be redshifted, which would mean we can determine whether time dilation applies at the quantum level—another important result. But if we try this and similar experiments at greater distances, we should see more intricate examples of quantum–gravitational interaction.

For one thing, the curvature of spacetime would affect the polarization of photons and hence any measurements of quantum entanglement. Near the Earth's surface, photons follow straight lines, which means that their horizontal and vertical polarization directions are fixed. This allows us to do clear-cut entanglement experiments in which measuring photon 1 as "horizontal" makes photon 2 yield "vertical," and vice-versa. But if two entangled photons follow different relativistic curved paths in space, the directions of their electric fields and polarizations would seem to depend on the details of the curvatures. In this way relativity becomes mixed into an innately quantum mechanical effect.

Polarized photons might also allow us to explore the hypothesis that spacetime itself is quantized—that is, not smooth as in general relativity, but a granular structure made of discrete Planck-sized cells. Evidence for this would be a huge step toward a theory of quantum gravity. The trouble is, we don't currently have any hope of probing these tiny units directly. This difficulty has inspired physicists to come up with an alternative—the Holometer at the Fermi National Accelerator Laboratory near Chicago. This sensitive device is designed to detect the "jitter" in laser beams as they are affected by the randomness inherent in a quantized spacetime. Some observers are skeptical that this will pay off, but data collection has started and could yield results relatively soon.

By contrast, an approach based on polarized photons in space came in 2010 from researchers at Imperial College, London. Each time a photon traverses a Planck cell, quantum randomness would slightly shift the direction of the photon's electric field. The accumulation of many quantum "kicks" as a photon travels a long distance through numerous cells would measurably change the polarization, indicating a granular spacetime. The distance would need to be billions of kilometers at least—the size of the Solar System—which is far beyond present quantum technology. But it's nice to dream big: the IQC paper classifies this experiment as "visionary." Whether that means "potentially ground-breaking" or "slightly off the reservation" is unclear.

Similarly visionary experiments could use entanglement to examine the history of the universe. In theory, as the cosmos grew from its tiny beginnings, the corresponding changes in spacetime curvature should have altered entanglement in ways that can be traced back to presently unknown details of cosmic development. Entanglement studies could even bear on an old cosmological question: are the physical laws that we have derived on Earth valid for the whole universe? Since the time of Copernicus, scientists and philosophers have considered this question from different perspectives, but have had little direct evidence to draw on. If entanglement proves to be truly infinite in scope, it could be the ultimate tool to glean answers from distant cosmic locations.

These wild schemes would have to use spacecraft operating far from Earth. But we've done that before: NASA's Voyager 1, launched in 1977 and now 20 billion kilometers away, is still out there in interstellar space, sending back data (as we noted earlier, space travel is pretty cheap once you've got going).

Voyager was never equipped to handle qubits or entangled photons, of course, but Singapore's Centre for Quantum Technologies has just tested a compact source of entangled photons suitable for spacecraft. There are huge technical challenges to overcome. It will be difficult to maintain entanglement over these distances. Even so, we *can* imagine putting such a source on a space probe and sending it off beyond the edge of the Solar System. Like Voyager, this vessel would have to travel for many years to reach a suitable distance. Its discoveries would lie a long way in our future.

More immediately, we can look forward to meaningful LEO satellite measurements, assuming that the underlying quantum network can be established. There are good reasons to think it can. The Chinese government is putting considerable resources into its quantum satellite program. Western capitalism is well-motivated to weigh in, too. A recent IQC market study asserts that in a world where data security is under heavy attack, QKD represents a "potentially lucrative business opportunity" because its fundamental quantum principles will always be proof against hackers. The study predicts that satellite-based QKD will form an important piece of a multibillion-dollar global quantum cryptography market. That ought to focus some energy.

Governments and financial institutions such as major banks will welcome the day when quantum satellites can securely transmit sensitive data. But the biggest winners of the quantum network will surely be researchers eager for new fundamental discoveries. And that day could well mark the first time in history when banks too big to fail test our best physical theories to destruction.

The Seductive Melody of the Strings

Review: *The Elegant Universe: Superstrings, Hidden Dimensions, and the Quest for the Ultimate Theory*, Brian Greene (Norton, New York, 1999).

Scientific literacy seems to be in short supply among the American public. Too many graduates of our best universities do not know the ABC's of science, such as basic facts about the solar system. Mathematics and physics may be the biggest problems; in fact, physicists learn to suppress a sigh when a new acquaintance says, almost inevitably, "physics was my worst subject in college." Yet people love hearing about exotic physical ideas, from quantum computing to black holes. That explains why books like Brian Greene's *The Elegant Universe* get written; the difficulty of conveying cutting-edge physics explains why such books are hard to write well.

Greene, a physicist at Columbia University, works in the area called string theory and does his best to present its abstract ideas to the general reader. The heart of the theory is this: physicists have long thought that the minuscule electrons, quarks, and so on that constitute matter at the smallest scale behave like mathematical points. String theory says no, there is a deeper structure: Each elementary particle is a particular mode of vibration of a minute oscillating string. To use a classical image, this change of perspective is like replacing Euclid's perfect geometric points with the harmoniously thrumming strings that Pythagoras related to the music of the spheres. That isn't all, for these elementary-particle strings vibrate in ten dimensions of spacetime. We see only four, because six of the dimensions are "rolled up" or "compacted" into a tiny format.

If you are unused to the dizzying heights of theoretical physics, if you trust visceral reactions more than equations, all this talk of strings is extremely strange. But we have learned to live with weirdness. After nearly a century, both quantum mechanics and the theory of relativity remain full of puzzling outcomes remote from ordinary human experience. Yet both theories work—they give predictions that can be verified by experiment.

So Greene makes a valid point when he urges his readers not to dismiss string theory just because it seems odd. Still, why would we want to substitute intricate vibrating systems for the purity of dimensionless points? This seems to violate an axiom stated by the English philosopher William of Occam in the 14th century, and still beloved by scientists: "What can be done with fewer assumptions is done in vain with more." Moreover, there is the embarrassing fact that this refined mathematical edifice has yet to produce clear predictions that can be experimentally confirmed. This shortcoming has engendered considerable controversy over the value of the approach, which has been criticized by the likes of the Nobel Laureate Sheldon Glashow, whose research has deeply probed elementary particles and their interactions. (Glashow proposed the existence of the "charm" quark, which was later found; and, with Steven Weinberg and Abdus Salam, he showed that electromagnetism and the weak nuclear force—two of the four fundamental interactions that make the universe work—are aspects of a single "electroweak" interaction.)

But as Greene explains it, the apparent complication is actually a great simplification, worth pursuing because the prize is so tempting. String theory seems to unite quantum physics, which describes the smallest scales, with general relativity, Einstein's theory of gravity that describes the biggest structures in the universe. Such unity has long been the brass ring that physicists strain to reach. Our best effort so far, the Standard Model, correctly describes the elementary particles and goes beyond electroweak theory to include the strong nuclear force. But it does not explain gravity, the fourth force. Now strings offer the possibility of including that as well to give, finally, a "Theory of Everything."

Greene's style in explaining string theory follows many of the prescriptions about science writing I give my students: a look at the history of the subject (it all began in 1968, with an insight made by a young theorist named Gabriele Veneziano); plentiful application of metaphor and analogy; and—often overlooked—the use of pictures when words fail, expressed here with especially handsome drawings. But to provide sufficient background to appreciate the potential power of the theory, Greene must cover quantum physics and general relativity as well as strings themselves. This is a lot of deep material, and its presentation is not helped by the dense writing and level of detail in much of this 450-page book. For the non-expert, less would have been a great deal more.

Nevertheless, Greene's belief in strings comes through as he writes about what it means to have a theory that awaits experimental verification, and his stake in that. Greene is correct in saying that opinion among even Nobel laureate physicists is turning toward giving strings a chance, if only because no better theory is available; still, he puts the rosiest possible spin on the prospects for confirmation. This is perfectly understandable, for Greene is one of the string theorists who, he writes, "know that they are taking a risk; that a lifetime of effort might result in an inconclusive outcome."

The risk seems enormous when we consider that the most direct test of string theory would require a particle accelerator at least the size of our galaxy. For the ultimate benefit of physics, and to save a great deal of theorizing from going to waste, let us hope that a more attainable test will show whether reality truly does dance to the music of these particular strings.

Time Examined and Time Experienced

"Time is nature's way to keep everything from happening all at once."

Though the meaning behind this quote could be taken literally, it reads like a joke. Thought to be originally written by the science-fiction author Ray Cummings in 1919, the phrase was used by American theoretical physicist John Wheeler in his chapter of the 1990 book *Complexity, Entropy and the Physics of Information.*

But Wheeler, who had a way with words, also knew how to be serious about time, and in 1986 he wrote, "Of all obstacles to a thoroughly penetrating account of existence, none looms up more dismayingly than 'time'...To uncover the deep and hidden connection between time and existence...is a task for the future."

The shift in tone from treating time as a joke to something deeper is a sign that we do not understand it, though, like fish in the sea, we are immersed in it. Even while expressing our ignorance about time, Wheeler himself had no choice but to self-referentially allude to one of its mysterious aspects—the future. And though he could not explain time, he reminded us that it has human as well as physical meaning when he wrote in that same chapter from 1990: "Heaven did not hand down the word 'time'. Man invented it...or as Einstein put it, *Time and space are modes by which we think, and not conditions in which we live.*"

Thinking about time

Wheeler and Einstein are not alone in pondering the nature of time. Philosophers and thinkers have done so for centuries, and no wonder: time both permeates all that we humans do and fascinates us when we consciously consider it. We endlessly speculate about its nature and about the possibilities of manipulating it and travelling through it. These were science-fiction themes even before H. G. Wells' classic *The Time Machine* (1895), and they still remain current, featuring in the 2014 film *Interstellar* and last year's Netflix series *Dark.* The late Ursula Le Guin's science-fiction novel of ideas *The Dispossessed* (1974) gives time special attention, with its physicist protagonist Shevek developing a "general field theory of time" to explain both its "sequency" (as Le Guin calls it) or linear evolution, and its relation to cyclic events like the orbiting of a planet around its sun or the repetitive sweep of the hands of an analogue clock.

In physics itself, scenarios involving relativistic wormholes hint at the possibility of time travel, while tachyons—hypothetical faster-than-light particles—could travel backward in time or send signals to the past. Although it seems unlikely that wormhole travel can be physically realized, and tachyons have never been detected, real particles going backward in time have meaning in the diagrams Richard Feynman invented to calculate the behavior of elementary particles. One of his insights in these useful representations is to show positrons as their antiparticles, electrons, traveling backward in time.

Despite dealing with such exotic notions, physicists have still not been able to produce a full theory of time. Lee Smolin of the Perimeter Institute for Theoretical Physics in Canada even argues in his 2013 book *Time Reborn* that physics is guilty of "expelling time" by not incorporating its fundamental reality. Nevertheless, physicists have long grappled with defining and using time as they try to explain the universe. Early in Isaac Newton's seminal *Principia* (1687), which laid out much of how physics functions today, he defined "absolute, true, and mathematical time" that "from its own nature flows equably without regard to anything external." Along with "absolute" and "immovable" space, to Newton absolute time formed a backdrop for dynamic behavior and physical reality that, his definitions imply, cannot be affected by human actions.

We abandoned the notion of absolute space in 1887 when Albert Michelson and Edward Morley determined the speed of light to high precision, with results that eliminated the ether, which was previously thought to be the space-filling entity against which motion should be measured. Absolute time was likewise abolished after Einstein re-analyzed what it means when we say two events happen at the same time, and then went on to derive special relativity. Now we know that time changes as measured by a moving observer and, according to general relativity, in a gravitational field: if time is a flow, its flow rate can be altered.

General relativity has also amplified the role of time in physics. Adding time to the three spatial dimensions through the term $c \times t$—the distance light moving at speed c covers in time t—gives a 4D spacetime manifold that concisely describes gravity and the universe. This has put time on a par with space, and relativity has also forced us to think more carefully about time. The "twin paradox"—in which a twin who rockets away from Earth at high speed returns younger than her stay-at-home sister—is an exercise in the variability of time and is also an example of time as a physical parameter with deep human effects.

Irreversibly forward

Other features of physical time may connect to our perception of it. The concept of entropy as a measure of disorder that always increases, at least in large systems over long times, has led to its label as the "arrow of time"—a physical progression that, unlike the reversible processes of classical mechanics, irreversibly points "forward" to define the apparent flow of time. That asymmetric one-way road seems integral to the human sense of time; as Feynman succinctly put it in his 1964 lecture at Cornell entitled "The Distinction of Past and Future." "We remember the past, we don't remember the future," he said. "We have a different kind of awareness about what might happen than we have about what most likely has happened."

But some subjective experiences of time differ from physical time. Pleasant events seem to pass quickly while unpleasant or boring ones invariably drag, though the measured elapsed time may be the same. These internal experiences depart from the objective measurement of time because of how our consciousness deals with it, as Wells understood. In *The Time Machine*, the Time Traveler who built that device explains to his friends, saying "There is no difference between time and any of the three dimensions of space except that our consciousness moves along it."

We do not know if consciousness moves along or through time, or simply provides a vantage point to observe time as it flows; but we do know that the brain does not produce a one-to-one correlation between its evaluation of time and temporal events in the outside world.

The philosopher and cognitive scientist Daniel Dennett from Tufts University in the U. S. has provided a model for this intricate behavior. In *Consciousness Explained* (1991) and elsewhere he proposes that the brain and consciousness operate under a "multiple drafts" approach. Instead of a central place in the brain that houses one's personhood and interprets sensory information—an idea that traces back to René Descartes in the 1600s—consciousness emerges from various functions occurring at different times in different parts of the brain. To bring together these neural events distributed in space and time, Dennett maintains we create a coherent internal narrative that is the "I" of a person, with personality, memory, and so on. The scattered behavior behind consciousness, he adds, guarantees that "the temporal order of subjective events is a product of the brain's interpretational processes, not a direct reflection of events making up those processes."

Dennett's counterintuitive model strongly contrasts with our internal sense of self and is not the only one offered by cognitive scientists. But regardless of the model, the relation between external time—whatever that really is—and our internal time is extraordinarily complex. Reporting on their work about how time and space are perceived by the brain, neuroscientists György Buzsáki and Rodolfo Llinás at New York University comment that "there is no doubt that the terms 'space' and 'time,' as well as other mental constructs, will be part of research for years to come."

Nature's clock

Our bodies too experience different aspects of time. Overlaid on our moment-to-moment responses to external events is the circadian rhythm—the approximately 24-hour cycle of physiological activity built into much of life on Earth, from people and animals to plants. In humans, it defines the periods of lowest body temperature, greatest alertness, sharpest rise in blood pressure, and deepest sleep. Though the circadian rhythm can be affected by external light and temperature, it arises from internal molecular mechanisms whose exploration led to the 2017 Nobel Prize in physiology or medicine for Jeffrey Hall, Michael Rosbash, and Michael Young.

The evolutionary benefit of the circadian rhythm is thought to be that it enables organisms to make the best use of light, food, and other resources depending on their availability at different times of the cycle. From the viewpoint of our reactions to physical time, this bodily rhythm shows that we respond to its cyclic nature along with its passage. It would be too much to say that the rotation of the Earth around its axis gave birth to time; but the regular alteration between light and dark must have impressed itself upon early humans and helped form our perception of time.

The passage of time hits us hardest as we age. Though now that we understand that time is variable, could we ever slow down the process? From measurements aboard aircraft and from the space satellites in the global positioning system, we know that relativity correctly predicts how time dilates with speed and gravitational field to make a traveler age more slowly. But at the comparatively low speeds and gravitational variation we can reach with today's technology, the changes are tiny compared to human lifetimes. NASA astronaut Scott Kelly spent over 11 months aboard the International Space Station starting in March 2015, but returned to Earth barely a few milliseconds younger than his identical twin brother astronaut Mark who remained on our planet.

However, the brothers participated in another kind of NASA twin study. With Mark as a control, they underwent extensive medical comparisons to determine how life in space—with its microgravity conditions, increased radiation, and psychological stress in unnatural surroundings—affects human health. The test revealed something unexpected as far as Scott's "telomeres" were concerned. Telomeres are DNA structures at the ends of chromosomes that protect them from damage but shrink as cells reproduce multiple times. This shrinkage seems to be a main cause of cellular ageing but Scott's telomeres actually lengthened. This observation is probably linked to the months he spent in space, rather than to the milliseconds of relativistic time change. Still, further exploration of differences in human ageing will surely occur in space when we achieve high speeds and changes in gravity, so we should keep in mind the relationship among human biology, time, space travel, and relativity.

Everyone's time

It is hard to envision a more multidisciplinary topic than time and how we perceive and react to it. The great contributions from physics are the development of relativity, which gives time a fuller, more flexible role in the universe; our understanding of time's arrow; and our remarkable ability to measure time with exquisite precision down to attoseconds while not knowing exactly what it is, with important outcomes for science and technology. A full picture of time, however, also needs neuroscience, biology, linguistics, anthropology, psychology, and even literature because of time's emotional impact. Feynman himself recognized this when he spoke at Cornell of "remorse and regret and hope" that "distinguish perfectly obviously the past and the future."

Literature too can powerfully distinguish past from future to help us understand what time means to humanity, as in F. Scott Fitzgerald's masterwork *The Great Gatsby* (1925). In the novel Jay Gatsby is a mysterious figure who started from poor origins. After gaining great wealth he tries to rekindle a deeply felt love affair from his earlier days and enter a new life, but tragically fails to transcend his past and realize his dream. Fitzgerald's haunting last words turn Gatsby's story into universal truths about how we strive to grasp time and how, at last, it inevitably slips through our fingers:

> Gatsby believed in the green light, the orgastic future that year by year recedes before us. It eluded us then, but that's no matter—tomorrow we will run faster, stretch out our arms farther...And one fine morning—
> So we beat on, boats against the current, borne back ceaselessly into the past.

The Six Elements: Visions of a Complex Universe

There are rosebushes in front of my house, and sometimes I step outside to stand among them, enjoying the look and scent of their blooms. I was raised in the city, and it still surprises me to see growing things spring from the earth. That may be why, as I pick up a handful of earth, I think how remarkable it is that its matter-of-fact grittiness leads to the intricate beauty of roses. It is natural for one then to think of Earth on a larger scale as one of the Four Elements of antiquity.

Earth, Air, Fire, and Water: All relate to growing things, and furthermore to life on our planet, to the structure of the planet itself, ultimately to all nature. The meaning of Earth, for instance, extends far beyond that of the rough, pragmatic material I can hold in my hand; it is the brute bulk of our world, expressed in massive geological formations, the substance of other worlds, the dense interiors of dead stars. It is precious minerals, useful ores from deep mines, and the exquisitely refined gemstones and solid materials we make from those raw substances.

With that last realization, I see that the archaic idea of Earth is still significant. Physicists study solids in laboratories, where it is easy to forget that they come from nature. It is at first startling, and then illuminating, to re-connect them to their origin, the element Earth. Water, Air, and Fire also have meaning in modern science. Water and Air correspond to liquids and gases, the other two great categories of matter; Fire can be interpreted as light, an energy that has always filled the universe, and as flame and heat.

Early philosophers considered two more categories: Void, the emptiness that contains matter, and Quintessence, a substance that forms the distant cosmos. These too have modern counterparts. Void is the vacuum beyond our atmosphere; as what physicists call the vacuum state, it is where the universe began. Quintessence corresponds to "dark matter"—the invisible material that is unlike anything on this planet and that makes up at least nine-tenths of the cosmos—and to "dark energy," the equally mysterious antigravity that is accelerating the expansion of the universe.

These ancient groupings are counterweights to the reductive style of modern science, which emphasizes nature's smallest units. This approach is highly successful, giving deep insights, for example, into the quantum nature of reality and the significance of DNA. However, it does not span the full picture that began 13 billion years ago with the Big Bang. The Six Elements—the Four plus Quintessence and Void—are universal categories that engender coherent visions of the physical universe, visions with artistic as well as scientific meaning, and the power to bridge those areas.

Two Greek thinkers in the 6th century BCE first conceived the natural Elements. To Anaximenes, everything was made of Air. Rarefied, it became Fire, which formed the Sun; condensed, it became Water and Earth. Thales took Water as primary; "All things are water," he wrote. A century later, the philosopher Heraclitus thought Fire was the basic substance.

A universe made from a single substance is elegantly simple, but a better model for the world's diversity came from the Greek philosopher Empedocles, born in Sicily around 495 BCE. A figure of mystical power, he was said to be able to raise the dead and later to have thrown himself into Mount Etna's crater to join the gods. His great accomplishment is the idea that Earth, Air, Fire, and Water form what lies around us, brought together or kept apart by Love and Strife. This brilliant scheme is broad enough to describe great swatches of the cosmos, yet sufficiently articulated to explain much. Bone, for instance, is Earth and Water blended together by the application of Fire, all in the proportion 1:1:2. Human vision occurs as Water modifies Fire; that is, the vitreous humor in the eye affects the visual ray, which in Empedocles' time was thought to come from the eye.

For all the power of the Four Elements, Aristotle later asked, "How could they account for the perfection of the heavens?" Accordingly he added a noble and incorruptible Fifth Element. Translated as *quinta essentia* in the Latin of medieval scholars, its name lives on in the word "quintessence." The ordinary world was made of Earth, Air, Fire, and Water; Quintessence made the distant cosmos.

One more essential category came from the Greek thinker Leucippus and his student Democritus, who in the 5th century BCE postulated that everything is made of irreducible atoms operating within the arena of space. "Nothing exists except atoms and empty space," wrote Democritus; "everything else is opinion." Or, as the Roman poet Lucretius put it 300 years later in his *De Rerum Natura* (On the Nature of Things):

> The nature of everything is dual-matter
> And void; or particles and space, wherein
> The former rest or move.

Void is the sixth category—seemingly colorless and inconsequential, but having, as with the "zero" in algebra, enormous consequences.

Today science sees the world differently. An "element" is not one of the Four or Six, but one of the hundred-odd types of atoms—hydrogen, oxygen, iron, and so on. Water is not elemental but a compound of hydrogen and oxygen. Atoms are not irreducible but made of electrons and quarks. Living things are made of cells, and cells have their own smaller components. This reductionism is essential to modern science, but the Elements provide a more unified comprehension of the universe.

Think again of the roses near my home. The plant life they represent underlies all earthly existence, including our own. A growing plant is supported by Earth, which also stands for any solid material. The plant needs carbon dioxide and nitrogen, and we ourselves need oxygen—which, along with other gases, forms our atmosphere, a complex form of Air. Water, H_2O, is also essential for life and covers nearly three-quarters of our planet. The Element Water also appears as sluggish petroleum and heavy, glinting mercury. Fire, as light from the Sun, floods our world. It too is essential, and life has evolved in response to it. If Fire is taken as flame and heat, it is also sunlight stored in coal, wood, oil, and gas and freed by burning.

Void, however, exists on our planet only in human-made vacuums, and Quintessence not at all. Both come into their own in the modern view that the universe began in the vacuum state and contains dark energy and dark matter. Aristotle was partly right in thinking that the celestial sphere is made of a different kind of material; still, the Element Earth appears in every planet and moon in our solar system, and in burned-out stars beyond it. Air is the hot chemical brew that is the atmosphere of Venus and the hydrogen widespread in space. Fire, taken as light, is the natural inhabitant of the cosmos and along with Void, the oldest. Created directly in the Big Bang, light's remnants still flash through the universe while new light glows from the stars.

Water, however, taken as Element or as the particular liquid H_2O, was once thought to be limited to our own planet. Now we know that H_2O has existed on Mars, that oceans of it may lie beneath a layer of ice on Jupiter's moon Europa and that petroleum-like lakes can be found on Saturn's moon Titan. Nevertheless, Water has the smallest cosmic presence of the original Four; contrary to Thales' philosophy, the universe is mostly dry.

Empedocles combined four Elements in varied proportions to form the world. Anaximenes envisioned a different evolution, in which Air thickened or thinned to form the parts of reality. Remarkably, both processes operate in the universe. There are countless examples of mixtures of Elements: For instance, ocean foam, made of Water and Air. Like Empedocles' Love and Strife, matter attracts or repels matter through four forces: the "strong" and "weak" forces within atomic nuclei; the electromagnetic force, which holds atoms and molecules together; and gravity, which works on every kind of matter.

Rarefaction and condensation occur as well. Soon after the Big Bang the Element Air first appeared as hydrogen gas. Under gravity, a cloud of hydrogen would become denser and hotter, until hydrogen nuclei would fuse into helium at a temperature of millions of degrees. That changed matter into energy, according to Einstein's relation $E = mc^2$, to form a glowing star. Later, helium nuclei would fuse into carbon. As the hydrogen became depleted and the star cooled and died, gravity would compress the carbon into a small, white-hot "white dwarf" of formidable density, weighing a ton per teaspoon. To Anaximenes, this conversion of hydrogen into surpassingly solid carbon would simply be Air condensing into Earth.

Condensation also formed our solar system. Five billion years ago, in a nebula made of hydrogen and dust, gravity accreted dust grains into pebbles, rocks and finally planets. Mercury, Venus, Earth, and Mars have cores made of various proportions of solid and liquid iron, surrounded by molten rock and a solid crust. These bodies also developed atmospheres and on Mars and on Earth, oceans. The other planets have rocky cores encased in solid, liquid, and gaseous hydrogen, helium, and other substances. In short, the planets condensed from Air and Earth into Earth, Air, Fire, and Water.

As an example of rarefaction, dense, massive stars can explode in a supernova, an inconceivable outpouring of energy like that from a trillion stars. Chinese astronomers observed one in 1054 AD. It tore a nameless star into rags of gas and dust, an enormous cloud of debris 6,000 light years away visible today as the Crab Nebula, its claw-like tendrils still spreading rapidly into space. The Crab, too, might please Anaximenes, for it is Earth transmuted into Air.

Some supernovas leave behind a hyper-dense core made of tightly packed neutrons. Its own ferocious gravity can squeeze the core down to a mathematical point—a distortion in spacetime through which ordinary matter disappears and from which light cannot escape. This is a black hole, predicted by Einstein's General Relativity and recently detected at the center of our own galaxy. In Elemental terms, a black hole is where Earth distorts Void and captures Fire, as matter at extreme density twists spacetime to trap light.

Although science now explains more than the early Greeks could, the Elements point to some remaining puzzles. For Water, we do not fully grasp how the molecules in liquids link together and how liquids flow in seemingly unpredictable turbulent swirls. This is also a great problem in gases, that is, Elemental Air. We understand the Element Earth in hard crystalline forms, as in salt or copper, but not so well in soft amorphous form, as in clay.

Of the Four, however, Fire as light is the most mysterious, perhaps because it is the least tangible. It is made of particle-like photons but also acts like a wave, in that paradoxical duality that has long puzzled physicists. Photons also undergo what Einstein called a "spooky" interaction: quantum entanglement. For reasons unknown, two photons can apparently communicate instantaneously no matter how far apart, as has been proven for distances near 100 miles. The ancient Greek thinkers struggled mightily with the nature of light, and that struggle still continues.

We have more evidence of Void and Quintessence than the Greeks did but do not fully understand them either. Void as vacuum was first discerned in the 1640s by Galileo's former laboratory assistant Evangelista Torricelli in his mercury barometers. Now we create vacuum at will and explore the vacuum of space. What is difficult to grasp is that random quantum fluctuations in nothingness led to the Big Bang, yet that seems to be the origin of the universe.

Quintessence became scientific in 1933, when the astronomer Fritz Zwicky postulated the existence of invisible or dark matter to explain how galaxies rotate in space. Dark energy was discovered in 1998, when astronomers found the universe to be expanding faster than predicted due to a universal repulsion that counteracts gravity. However, we do not yet know much more about these facets of Quintessence than Aristotle did.

In addition to providing a structure for scientific thought, the organizing and metaphorical power of the Elements has inspired artists in different eras. Sometimes the Elements appear allegorically, as in Nicolas Poussin's painting *Cephalus and Aurora* (1627–1630). It shows Cephalus rejecting the love of Aurora, goddess of the dawn, out of loyalty to his wife. The scene is embellished by the Four Elements, represented by the Earth goddess Gaia; the winged horse Pegasus, for Air; the sun god Apollo, for Fire; and the sea god Oceanus, for Water.

Other artists portrayed the Elements realistically based on observation of the world, the common basis of science and art. Leonardo da Vinci often rendered Water in motion, as in one painting of his *Deluge* series (c. 1510) that illustrates the turmoil caused by rocks crashing into water. Later that century, Caravaggio painted his famously intense lighting effects. They influenced Georges de La Tour, a close observer of Fire. His *St. Sebastian Attended by St. Irene* (c. 1649) shows the flame from a torch as if the artist has frozen its dynamic twisting forms with a camera. Void was expressed by the 18th-century English painter Joseph Wright. His *Experiment on a Bird in the Air Pump* (1768) shows a lecturer demonstrating vacuum by pumping the air from a chamber containing a white cockatoo that is about to expire.

Some artists use the Elements directly. Ground up and mixed with a carrier, Earth has been a source of color pigments since the prehistoric cave-painters. Now it is the medium in Robert Smithson's massive earthwork *Spiral Jetty* (1970) on Utah's Great Salt Lake. Air and Water also appear. One of Marcel Duchamp's ready-mades from 1919 is a small glass ampoule filled with Parisian air. Decades later, in his installation *Tramway* (1976), the radical German artist Joseph Beuys filled an iron pipe with water from beneath the work's site in

Venice. Other artists exploit moving water. Fire has flourished in art at least since 1963, when Dan Flavin made glowing sculptures from fluorescent lamps. Later, James Turrell filled space with seemingly tangible light. Void appears in *Development of a Bottle in Space* (1912) by the futurist Umberto Boccioni, which shows solid matter flowing into space that flows back into solid matter; and in Aleksandr Archipenko's *Walking Woman* (1912) and *Woman Combing Her Hair* (1915), which use open areas to represent a woman's face and body within solid bronze forms.

The Belgian surrealist René Magritte created his own Elements. In *The Six Elements* (1928), panels set in an irregular frame show fire; sky and clouds; deep woods; a naked woman's torso; the facade of a house; and bells from a horse's harness floating over a lead curtain, a recurrent theme for Magritte. These scenes appear in other works, such as one version of *The Empty Mask* (1928), except that there a paper cutout replaces the torso. In each painting only two or three of the six fit into the original Four, making the remaining ones strangely evocative.

Another version of *The Empty Mask* (1928) shows four elements, each denoted by a word or phrase rather than an image. Magritte's characteristic use of words in his art presents an ingenious way to sidestep the difficulty of portraying "Void" by simply painting the word "vide" inside a frame. But whatever his choices for the Elements and their representations, what is important is the act of choosing; as the noted art critic and historian A. M. Hammacher has put it, by his selection Magritte "regroup[s] the isolated elements of life . . . [he] dissolves the untidy picture of the ordinary world into a number of vital factors" [1].

This analysis of an artist's thinking aptly describes the continuing power of the categories defined by those Greek thinkers to impose sensible order on the world and our understanding of it. The Elements, whether the Four or the Six, have a way of uniting art and science—and of encouraging us to see distant worlds in the petals of a rose.

Reference

1. Hammacher, A. M. *René Magritte* (Harry N. Abrams, New York, 1995), p. 98.

Stealth Science

Review, *Out of the Crystal Maze: Chapters from the History of Solid-State Physics*, Lillian Hoddeson et al., editors. Oxford University Press, 1992.

Have you ever heard of the "step back" syndrome? If you are a physicist, you no doubt encounter it regularly. I see it most often at cocktail parties, in exchanges like this:

INQUISITIVE FELLOW GUEST: And what do you do?

ME: I'm a physicist.

IFG [*Takes involuntary step back, remains silent for a moment*]: Oh . . . that was my worst subject in college. [*Further silence.*]

Smiling nervously, I remark that many people have survived a bad physics experience and gone on to lead happy, productive lives. But my attempt at gentle irony just delays the inevitable.

IFG: What kind of physics do you do?

ME: Solid-state physics—you know, the study of solid matter.

IFG: Oh. [*Prolonged silence.*]

That reaction always surprises me. If I were to say "nuclear physics" or "astrophysics," my inquisitor would nod sagely at the mention of a familiar scientific specialty—yet who has ever seen an atomic nucleus or a black hole? Solid-state physics, unknown as it seems to be, deals with a reality anyone can know and touch. And it is central to 20th-century science. Its problems have stimulated sophisticated scientific thought, such as Einstein's early reflections about the thermal properties of solids; its breakthroughs have earned Nobel Prizes, such as the one awarded in 1987 for the discovery of high-temperature superconductivity. A huge plurality of American physicists works in the area, far more than in elementary-particle physics or in any other subfield. Every March thousands of them troop off to the annual meeting of the American Physical Society to hear their colleagues present thousands of research papers. Research in solids pervades many federal laboratories and underlies much of the American economy.

And yet that important enterprise remains a kind of "stealth science." Rarely has it been treated in print as cohesive scientific area, and never before with the sweep of *Out the Crystal Maze*.

Maybe solid-state physics is unappreciated because it seems less compelling than the search for cosmic roots. Both cosmology, which examines the origin and fate of the universe, and elementary-particle physics, which seeks its ultimate components, traffic in concepts unparalleled in ordinary experience: unimaginably vast expanse of space and time; infinitesimal entities, such as quarks, that no one will ever grasp or hold. Meanwhile, working between the extremes of duration ad size, solid-state physicists hold the material world at arm's length to scrutinize the substance of their own planet. Their object of study, solid matter, is literally mundane. Yet as a component of the big picture, it is also exceedingly rare. Most of the cosmos appears as gas, flaming plasmas, or inscrutable "dark matter;" only a minute fraction of the universal stuff is bound up in planets, dead stars or microscopic particles of interstellar dust.

Nevertheless, the world at our fingertips is rooted in cosmic events. Out of the Big Bang came electrons and quarks, and the quarks assembled themselves into protons and neutrons. About 300,000 years later, those particles first united to form atoms, which went on to combine into the marvelous intricacy of matter. If the idea of elementary particles fascinates through its categorical simplicity, the characteristics of solids, liquids, and gases impress through their overwhelming diversity. Of those three categories, gases are the simplest to analyze, whereas the "condensed matter" of liquids and solids requires more intensive study. The range of solids, natural and human-made, is particularly rich. The world is blessed with the fragile clarity of glass, the useful magnetic properties of iron and electrical properties of gallium arsenide, the luminous beauty of ruby, the tough resilience of plastic.

Familiarity dulls wonder, and it is easy to forget that such varied characteristics represent profound properties of the physical universe. A solid is made up of atoms held billionths of a meter apart by electromagnetic forces. Even a barely visible speck of matter is a busy little cosmos populated by an astronomical number of atoms. In some solids the atoms are randomly arranged, but in crystalline materials such as metals and minerals—the heart of solid-state physics—they lie in regular arrays. Their intricate geometry determines how the atomic nuclei and their associated electrons interact. It explains why sapphire is blue, why silicon is a semiconductor. But within those arrays there lurks a mystery, a

Minotaur in the crystal labyrinth. Through their minute arcades, electrons move not like particles but like waves, exhibiting the wave–particle duality that marks the dark side of the most paradoxical theory in science: quantum mechanics. At the same time, the study of solids confirms the validity of the quantum theory in deep and surprising ways.

It takes a big book to tackle such a sprawling and complex area. *Out of the Crystal Maze* fills the bill, with 713 pages weighing in at three and a half pounds. Its four editor-contributors are veterans of what they suspect is the "largest international research project ever done in the history of modern science," the roots of which extend back to the 1970s. The project involved historical research teams based in France, Germany, Great Britain, and the United States. Nine other scientists and historians of science also helped write the book. Fully aware that even such a massive enterprise could not cover everything, the editors chose to end the historical narrative in 1960 and to treat only a selection of scientific topics—a pioneering effort that should stimulate others.

I tested the book's coverage by my own subjective method, checking to see whether my dissertation adviser is named in the index. He is, but more important is that the chapters are well chosen to represent key subject areas and their times. The first three trace the intellectual history of solids. The pre-quantum era covers millennia, from ancient human manipulations of matter until the early 20th century, when new experiments and theories illuminated the microscopic nature of solids. That outlook explained some solid state puzzles—the origins of magnetism in iron, the splitting of an image as seen through transparent crystals—but it was incomplete without quantum mechanics.

The existence of quantum phenomena was first suspected in 1900, when the idea that energy can take on only certain discrete, quantized values was successfully invoked to explain the observed spectrums of hot bodies. By 1926 the quantum behavior of a single hydrogen atom could be accurately described. Also in 1926 came the first quantum-mechanical ideas of how to deal with electrons in carload lots, which led to a correct theory of solids. In that era, which extends to the present day, quantum theory has been honed until it cuts through many of the knots in the understanding of solids.

The importance of solid materials in human culture is reflected in the names of the Stone, Bronze, and Iron ages. In the 19th century, new industrial technologies required new materials: various kinds of steel, exotic substances for the filaments of incandescent lamps, magnetic alloys for telephones and electrical devices. Engineering sciences such as metallurgy met some of those needs.

A true science of solids, however, began with the discovery of the electron by the English physicist J. J. Thomson in 1897—a seminal discovery that also gave birth to the science of elementary particles. Just three years later the German physicist Paul Drude theorized that metals, which are good electrical conductors, are filled with electrons that move freely, like the atoms in a gas. Drude's free-electron picture accounted for Ohm's law, the basic relation among voltage, electric current, and electrical resistance. But Drude's theory failed when it was applied to the thermal property called specific heat, the energy required to raise the temperature of one gram of metal (or any material) by one degree Celsius. To classically trained physicists, it seemed obvious that the septillions of electrons in a few kilograms of metal should absorb energy as the metal was heated. And if so, a metal would have a greater specific heat than an electrical insulator, which lacks free electrons. Yet experiments showed no such difference; something was wrong with the classical picture.

The solution to the conundrum, it turned out, was to recognize that a swarm of electrons follows quantum rules. One of those rules, which the Austrian physicist Wolfgang Pauli articulated in 1925, requires that no two electrons simultaneously have identical values for a set of characteristics that include energy and a quantum property called spin. Spin is related to the magnetic field associated with an electron, and that field is quantized in the sense that it can point in only one of two directions—generally labeled Up and Down. It is as if the electron were a minute sphere with its charge distributed throughout its volume, spinning in such a way that its electric charge circulates either clockwise or counterclockwise. According to Pauli's exclusion principle, two electrons with opposite spins can have the same energy, but a third electron added to the ensemble would necessarily match the spin direction of one of the first two. Hence additional electrons must take on different energies, pair by pair.

Pauli's fundamental restriction led the physicists Enrico Fermi and P. A. M. Dirac—two future Nobel laureates who decisively influenced 20th-century physics—to derive in 1926 a new quantum statistics of electrons: a new way of describing electrons in the aggregate. According to Fermi–Dirac statistics, electrons in a group act like marbles packed into a narrow glass beaker. No two marbles can occupy the same space, and so the column grows taller as marbles are added. Only the top layer of marbles can easily be raised to new heights, because a marble buried in the stack is surrounded by neighbors that would also have to be shifted. Similarly, each succeeding pair of electrons must settle at a higher energy than its predecessors as it is added to the group. The entire collection of electrons occupies a range of energy values extending from zero to a maximum called the Fermi energy. Of all the electrons in a metal, only a negligible fraction near the Fermi energy participate in thermal events, which explains why electrons do not contribute to specific heat.

That result was a triumph of the quantum theory. Still lacking, however, were details about the atomic arcades within which, according to the theory, electronic waves must reverberate. Since 1784, when the Abbé René Just Haüy published his *Essai d'une Théorie sur la Structure des Cristaux*, it had been known that the external shape of a crystal reflects its internal structure. Haüy's insight came when he accidentally dropped a crystal of calcite—a compound of calcium, carbon, and oxygen that occurs in limestone and marble, stalactites and stalagmites, alabaster, and lapis lazuli. He noted that the broken shards were miniatures of the original crystal—a tilted cube, or rhombohedron. In a beautiful drawing that has been reproduced many times, Haüy showed how small cubes could be stacked to form a macroscopic crystal, like identical children's blocks interlocked into a structure.

It is now known that the basic crystalline unit is made up of atoms. They are too small and too close together to be seen with the comparatively long wavelengths of visible light. But the small wavelengths of X-rays, discovered by the German physicist Wilhelm Conrad Roentgen in 1895, can resolve crystal structure at the atomic level. In 1912 they first showed how atoms form a repetitive lattice in a crystal. (This part of the tale reminds me of how small the physics community really is, and how much is passed from professor to student. Two of my undergraduate teachers at the Polytechnic Institute of Brooklyn, Paul Peter Ewald and Isidor Fankuchen, both now deceased, were among the pioneering X-ray analysts of crystals.)

The lattice determines the mechanical properties of the solid. Perhaps even more important, its virtually endless arcades of recurrent atoms modify the electronic waves, in ways analogous to the ways an ordinary hallway leads to echoes and reverberations of sound. In the 67 years since that was established, the theory of electronic waves in a solid has grown into a full "band theory," which states that electrons in the lattice occupy continuous bands of energies, not the discrete levels of a single atom. The different bands, along with the "band gaps" that separate them, define the behavior of the electrons and determine many of the properties of the macroscopic material of which they are a part.

The next five chapters of *Out of the Crystal Maze* trace how those elegant theories came to illuminate the properties of actual metals and insulators, semiconductors, superconductors, and magnetic materials. Reality intrudes in the form of crystalline defects, sites where the perfect lattice is spoiled by missing or disordered atoms. A line of displaced atoms in a metallic lattice is a miniature geological fault line. In sufficient numbers, such dislocations determine how the metal deforms under stress. Foreign atoms also mar crystalline perfection. Sometimes that is desirable, as when natural impurities color the compound aluminum oxide to make red rubies or blue sapphires. Alien atoms can even be added deliberately, as is done to control the electrical properties of semiconductors.

The "semi" in semiconductors means that such materials have electrical properties midway between those of metals and those of insulators. Band theory explains why. Whereas metals have many free electrons, and insulators virtually none, semiconductors carry free electrons in intermediate numbers that vary with impurities and temperature. That variability leads to electrical properties useful in electronic devices as well as to optical properties that make some semiconductors suitable for use in lasers and light sensors. Band theory has reached a high pitch of sophistication; it is now possible to compute the electronic properties of both metals and semiconductors, starting from the details of the atomic lattice.

Band theory fails, however, to account for certain collective effects, namely, the ones that arise when the electrons in a solid influence one another. Such effects explain the puzzle of superconductivity, in which the electrical resistance of certain metals can suddenly fall to zero at extremely low temperatures. The effect was first seen in 1911, and its elucidation earned a Nobel Prize for John Bardeen, Leon H. Cooper, and John Robert Schrieffer in 1972. The electric charge of an electron in a superconductor pushes and pulls on the lattice, thereby distorting it; the distortion affects a second, distant electron, which thus begins to move in step with the first. The electron pairs act as a single, bound system, much like a double star, in which the motion of each partner affects that of the other.

Such pairs, in the aggregate, have properties radically different from those of free electrons in a solid. The pairs are not subject to the Fermi–Dirac restrictions on sharing energy values. Instead they follow a different set of statistical rules derived in 1924 and 1925 by Einstein and the Indian physicist Satyendra Nath Bose. In the Bose–Einstein world, the pairs act like marbles in a spacious pan rather than in a narrow beaker. No matter how many marbles are added, they spread out in a pool one marble deep—that is, the electron pairs all have the same energy. It takes a relatively large amount of energy to detach any single electron from the condensed mass, because that would upset the energy state of the entire aggregate. Hence the electrons flow through a superconductor without collisions, which would have to be quite violent to tear the pairs apart.

Superconductivity is a human-scale manifestation of seemingly unfathomable quantum rules, which physicists still seek to understand. Their search continues to make front-page news; as I write this review, investigators have just announced that they have made the first Bose–Einstein condensate, linking a set of free atoms into a new, coherent state of matter. The scientific depth in solid-state physics comes from its relation to the quantum—yet its societal aspects are equally significant, for it straddles the ideas of "pure" and "applied" science in ways that cosmology and astrophysics do not. The first chapter in *Out of the Crystal Maze* describes how solid-state physics, technology, and industry were interwoven in the 19th century; the last chapter relates how the scattered study of solids became today's directed research discipline, which continues to value both applied and academic studies. Important work takes place in industrial laboratories as well as on university campuses; the science of solids underlies the computer and communications industries, which may make our era memorable as the "semiconductor age."

The combined goals in solid-state physics carry special weight these days, when science is continually called upon to justify its costs to society. A great nation, it seems to me, must support the fundamental scientific quests as well as those of artists, explorers, and philosophers; yet it must also value applied science, which can meet pragmatic human needs. The costly Superconducting Super Collider (SSC), designed for research on elementary particles, failed partly because it was justifiable mainly as an intellectual enterprise, with little prospect of generating useful outcomes. The science of solids, however, can maintain the essential balance between pure knowledge and the commitment of technology to human service.

Other cultural characteristics of solid-state physics are also identified in the last chapter of the book. The study of solids tends to operate as "little" science, carried out by individuals or small teams rather than by large research teams wielding enormous apparatus. Individual creativity was an incentive for many of the pioneers of solids. The book quotes the physicist Frederick Seitz, former president of the National Academy of Sciences and of Rockefeller University, as calling solid-state studies a field in which "a person could still be himself and not a slave to a machine"—exactly what appealed to me when I chose the area in graduate school. And its practitioners accept that real-world materials cannot always be described by the rigorous mathematics that characterizes other branches of physics. They apply tools such as computer simulation and do not disdain intuitive insight. That attitude may contribute to the general study of intricate real systems, from economic markets to biological organisms, which some call the science of complexity.

Every part of solid-state science—even what is not represented in *Out of the Crystal Marc,* such as the physics of glass and other amorphous materials—returns to the real and the tangible. Nothing, it might seem, could be further from the stars, but in fact the hard reality of solids is essential in the search for cosmic roots. Exquisitely shaped metal and glass form the mirrors and lenses that bring the sky within reach; sensitive semiconductor devices detect celestial light; superconducting magnets guide electrons and protons in giant particle accelerators; and a substrate of humming computers full of silicon chips underlies virtually all of modern science.

Apart from that supportive role, the science of the minuscule carries its own sense of wonder. William Blake expressed it when he wrote of seeing "a world in a grain of sand," of holding "infinity in the palm of your hand." Not long before Blake composed those lines, Abbé Haüy had actually held a crystalline cosmos. He had seen it shatter to bits, yet each piece still contained a complex, ordered atomic universe. Now materials scientists have learned to form miniature worlds. One contemporary investigator, Richard E. Slusher of AT&T Bell Laboratories, assembles layers of semiconductors into lasers so small that thousands fit into a square inch of space. He calls the minute structures nanocathedrals, reminding one that the small is as marvelous as the huge and distant.

An appeal to the pleasures of the small might not satisfy my fellow party guest, who might want to hear other reasons to study solids. I suppose I could simply hand over a copy of *Out of the Crystal Maze*. But big as it is, that would be impractical, even in paperback. Or I could make the point more sharply. When information is stored in RAM and on ROM chips, on diskettes and hard drives, I might point out, the storage media are made of semiconductors and magnetic materials, products of the very solid-state research described in *Out of the Crystal Maze*. The book's 400,000 words could be compressed into a single silicon chip, or onto two floppy diskettes. A book, too, is a universe in the palm of the hand, a universe that shrinks dramatically when it is stored in solids as strings of bits—a powerful illustration of the command of the atomic world that has come from contemplating the solid state.

Froth with Meaning

I have always liked coffee but my delight took on a new dimension when I began drinking espresso and cappuccino. Now my spirits lift every morning as I grind the rich, dark beans, measure the fragrant powder into the espresso machine, and watch the drip. Then comes the best part: I plunge the steam pipe into milk, making bubbles that cling together closely to form the magic froth that turns espresso into cappuccino. Sometimes the bubbles are small and densely packed, making a stiff, freestanding foam, like well-beaten egg whites. Sometimes they are large and fragile, forming tenuous, milky suds that soon collapse.

I enjoy foam while I drink my coffee and yet, with equal pleasure, I could bring it into my physics laboratory for serious study. Over the years I have conducted and published research on every kind of matter, from exotic metals to crystalline sapphire, from liquids and gases to complex biological molecules. Yet few of the systems I have studied present the scientific challenges of foam.

It may be startling to realize that such an airy substance carries scientific weight. But the universe is awash in foam. On the planetary scale there is the foam of whitecaps, which covers millions of square miles on the world's oceans and influences the earth's climate. Pumice, a kind of foamy rock spewed from volcanoes, carries clues to the geologic history of the earth. On a cosmic scale the billions of galaxies that make up the universe are arranged as if they lay on the surfaces of immense bubbles within a gargantuan foam.

To show the students in my astronomy class the foamy arrangement of galaxies in the universe, I make a simple model of the cosmos: a small, transparent plastic bottle half-filled with ordinary tap water and a dash of liquid soap. Shake the bottle and the clear liquid turns into a mass of bubbles, each surrounded by soapy water. The universe itself is structured that way—except that each bubble is a stupendous volume of space stretching across hundreds of millions of light-years, and the stuff between the bubbles is made up of billions of galaxies, each with hundreds of billions of stars.

My plastic-bottle universe works equally well as a laboratory model of how earthly foam is born. First, the model demonstrates that you need more than pure gas and pure liquid to make foam. Yes, the surface tension of the water provides the skin surrounding each bubble, and bubbles form when you shake the bottle; but omit the soap and the bubbles die the instant you stop shaking, too unstable to yield even a short-lived froth. Only by adding that bit of liquid soap, and shaking well, do you get bubbles that are robust enough to froth. Why does adding the soap bring about such a powerful change?

The soap molecules, which are released when soap dissolves in water, are long hydrocarbon chains whose two ends have quite different properties. One end is electromagnetically attracted to water, whereas the other end tends to reject water. Such a material is called a surfactant, a name derived from surface-active agent. The asymmetrical attractions between a surfactant molecule and water enable the surface tension of the water to vary across the curvature of a bubble. That, in turn, enables a bubble to adjust to any force, such as gravity, that would otherwise destroy it. In addition, the soap molecules interlock with one another and with the surrounding water molecules, strengthening the surface of the bubble. To put it another way, they create a skin from which water drains only slowly, extending the lifetime of the skin and the bubble.

The result is that when you shake a mixture of air, water, and surfactant, you get relatively stable bubbles of air whose skin is flexible enough to adjust to outside stresses—a genuinely radical transformation. Transparent air and water are transmuted into an opaque white mass. The slug of liquid that you feel sloshing back and forth changes as it becomes mingled with air, to be replaced by a mass that hardly moves as you shake and feels. . .well, frothier. What could make it clearer that a foam is utterly different from its constituents?

But what, exactly, is the nature of the difference? Unfortunately, there is no single theory of foam; for all its frothiness, foam is a surprisingly complex state of matter. It is not exclusively a solid, a liquid, or a gas but is made of bubbles or cells of gas within a liquid or a solid, and combines the characteristics of its components in novel ways. The physicist David A. Weitz of Harvard University has studied foam for 10 years and has called it the "neglected material"—neglected, that is, by scientists.

Physical scientists generally analyze the universe by breaking it into progressively smaller bits: first molecules and atoms; then subatomic protons, neutrons, and electrons; then (in the case of the protons and neutrons) their constituents, the quarks. But to grasp the surrounding world, one also needs to understand how the atomic building blocks relate to the materials of common sense. The connections are easiest to trace for a pure crystalline solid, such as salt or diamond, whose atoms occur in specific, repeating patterns, just as a huge hotel is built up from the repeating units of its rooms.

But the real world also includes clay, wood, and other non-crystalline solids; liquids, which are harder to understand than solids; and, most complicated of all, various combinations of solids, liquids, and gases. Foams are combinations, of course, but a variety of other combinations exist as well: in emulsions such as milk, bubbles of one liquid float in another liquid without mixing. In colloids such as gelatin, minute specks of a solid substance are distributed throughout another material, typically a liquid. In each of these combinations, including foam, the inclusions are arranged randomly in the surrounding medium. That usually makes the materials easy to deform: a bowl of gelatin, for instance, readily sets into the shape of its container. Although a few foams, such as pumice and certain plastics, set into stiff, solid forms, foams and the other combined systems are called soft matter—neither flowing freely, as a true liquid does, nor taking on the hard, definite shape of a rigid solid such as a diamond.

New properties arise when substances are mingled. Think about a foam such as soapsuds, which is made of liquid and gas. The foam is neither fully liquid nor fully gaseous; it flows, but not the way a liquid does, and it does not dissipate the way a gas does. Whereas its components are stable, the foam lives for only a short while. And though neither air nor water sticks to the hand in any great quantity, if you scoop soapsuds into your palm and turn your hand over, the suds will remain in place. In most respects, a foam is entirely unlike the substances that make it up.

A fundamental step in the study of foam is the examination of its bubbles. This is not easily done, because light does not readily penetrate a foam. When a ray of light passes from air into plain water, it bends slightly, then continues in an undisturbed straight line. In a foam, however, the ray soon encounters a bubble. Part of the light is reflected at the surface of the bubble, and part of it continues through the bubble. The new rays, traveling in different directions, encounter other bubbles, which again split each ray into two directions. The result is that light rays wend through a foam like balls in a three-dimensional pinball machine, bouncing off bumper after bumper to trace complex, nearly unpredictable trajectories. Although some light may survive to emerge from the far side of the foam, all that can be detected is an undifferentiated white glow.

One ingenious method to overcome that limitation modernizes a simple technique employed by the English physicist Charles Vernon Boys at the end of the 19th century. Boys made a box-shaped cell with transparent glass sides, much like an aquarium tank but measuring a mere half inch from front to back. When he poured a solution of soapsuds into the cell, only a layer or two of bubbles could fit into the narrow space. Because those layers were not obscured by other layers, Boys could readily examine the size and shape of the bubbles that fanned. Contemporary investigators have improved on the idea by combining a glass cell with modern office technology: a cell filled with foam is placed horizontally on an ordinary photocopying machine. By making a rapid-fire sequence of images with the copier, it is possible to study the evolution of the foam with time.

Another innovation takes the glass cell a step further. The physicist Florence Elias of the University of Paris and her colleagues have come up with a kind of magnetic soapsuds that enables them to explore the dynamics of a foam. Instead of studying bubbles of air formed in water mixed with molecules of soap, they study bubbles of oil (which, like air, is less dense than water) formed in water that is laced with small magnetic particles. The magnetic forces link the particles together, just as interatomic forces link the soap molecules in water, to act as surfactants. The resultant films define bubbles of oil in the water. When the mixture is placed between glass plates, it forms the familiar polyhedral cells seen in soapsuds. The difference, though, is that the novel magnetic "suds" can be stretched and pulled simply by placing a magnet nearby, which provides an easy but powerful way to study how foams respond to varying conditions.

Boys's original technique and its updated descendants, however, suffer from a common shortcoming: the investigator can study only a thin slice through a foam and cannot observe its deep interior, far from its external boundaries. What investigators need is a way to look through a foam and analyze it in its full three-dimensionality.

One useful, albeit still limited, approach is to probe the foam with light. It turns out that even though all semblance of an image is lost when light is scattered in a foam, the scattered light of a laser beam that emerges after passing through a foam carries information about the many bubbles the beam encountered deep inside the foam. Recovering that information from the beam is called diffusing-wave spectroscopy (DWS), because the analysis is applied to scattered, or diffused light. DWS was first carried out on a foam in 1991 by Weitz and the physicists Douglas J. Durian and David J. Pine, all of whom were then working at the Exxon Research and Engineering Company in Annandale, New Jersey.

To test the method, the three investigators examined a common foam, a commercial shaving cream fresh from the can. They put some of it into a glass cell, shone a blue laser beam through one face of the cell and measured the fraction of laser light that emerged after traversing the foam. What gave meaning to such a straightforward measurement was that the investigators had derived a simple and powerful mathematical relation between the emerging light and the properties of the foam: the amount of light that penetrates a foam depends on the average size of the bubbles. With that knowledge they could, in effect, watch the shaving cream coarsen: as time went on, the bubbles became larger, on average, and the foam transmitted more light.

Unlike Boys's glass cell, the laser probe makes it possible for investigators to examine the coarsening of the bubbles throughout the foam, even deep inside it. But it has also revealed something new. The experimenters noted fluctuations in the intensity of the transmitted light, indicating that something was rapidly changing in the foam. Upon inspecting the bubbles through a microscope, the physicists eventually observed what they call rearrangement events, in which several neighboring bubbles suddenly and simultaneously shift their positions. Weitz and his coworkers have surmised that as the bubble coarsening continues and the bubbles change size, stresses develop that make it easier for the bubbles to move than to take on new shapes.

As valuable as the DWS method has proven to be, laser light can probe the properties of foam only along the path of the beam, not as a complete three-dimensional array. One promising technology is magnetic resonance imaging (MRI), the diagnostic procedure that enables physicians to look inside a living body. MRI machines explore the body through the magnetic properties of atomic nuclei, such as the hydrogen nuclei that are found in every molecule of water. By measuring the positions and concentrations of the hydrogen nuclei, an MRI system can calculate the shapes and positions of differentiated structures inside an otherwise opaque watery volume such as a living body, and display those structures as an image. An MRI instrument can look equally well inside an opaque foam, provided the foam is partly made up of magnetically responsive nuclei.

The first published three-dimensional portrait of a foam was made in 1995 by investigators from the University of Notre Dame in South Bend, Indiana. The foam in question was made of gelatin mixed with water, and a commercial MRI machine generated a striking three-dimensional image of its multitude of bubbles. Since then, MRI instruments have made it possible for researchers to observe the size of bubbles as a foam ages. But because magnetic resonance imaging takes so much time, the method cannot follow the rapid changes characteristic of a flowing foam, which optical systems such as DWS can readily capture. For that reason, both methods remain valuable and complementary.

Because the study of foam is hindered by the very qualities of the subject matter, computers can be invaluable to investigators in sidestepping some of the laboratory limitations. One typical computational technique is to create a virtual foam that simulates a real one. Douglas Durian and his wife, the chemist Andrea J. Liu, both now at the University of California, Los Angeles, recently ran a computer simulation to explore foam in motion. Even with a powerful computer, however, the calculations are demanding, so Durian and Liu's virtual foam was limited to several hundred bubbles moving in two dimensions. The computer calculated the size and position of the bubbles as they responded to varying levels of stress and drew a picture of the foam in each case. According to Durian, the snapshots show that when a foam flows, large numbers of tightly packed bubbles suddenly snap from one configuration into another, like rocks on a hillside tumbling over one another as an avalanche roars down the slope.

Many properties of a foam ultimately depend on the effects of surfactants such as soap, which enable the bubbles to adjust to external forces. But soap is by no means the only surfactant. Seawater, for instance, incorporates a rich stew of molecules that play such a role. That is why the sea sustains foam, whereas freshwater does not—unless it carries additional surfactant molecules from polluting compounds.

Perhaps the relative composition of surfactants does not naturally spring to mind when one is contemplating the health of the crema atop a cup of espresso. One would rather test it with a teaspoon of sugar or some chocolate shavings. To the experienced espresso drinker, the appearance of a substantial crema shows that all is well with the coffee and its brewing. If the foam can support the weight of the sugar, it is a sign of a well-made espresso.

But crema has an importance that goes beyond its pleasing look and its value as an indicator of the brew: like the head on a beer, it affects the overall flavor of the drink. In a recent analysis at the University of Aveiro in Portugal, the food chemist Manuel A. Coimbra, his student Fernand M. Nunes, and their colleagues brewed espresso from coffee grown in Brazil and Uganda. They subjected each brew to a battery of tests, which showed espresso to be an intricate mixture of sugars, fats, proteins, and carbohydrates, including polysaccharides (large, complex polymers made up of units of sugar). According to the analysis, the proteins are the surfactants that make the crema. The polysaccharides increase the viscosity of the coffee so that it drains slowly from the foam. The crema traps the volatile compounds that carry aroma, thereby slowing down their release. That slowdown extends the pleasure of the drink by delivering its flavor in what Coimbra calls a "dosed" fashion—it is released over time, much as a steady stream of medication emerges from a time-release capsule.

The Portuguese investigators found that heavily roasted coffee beans produced foam of a rich golden-brown color, and also gave the greatest amount of foam. The foam lasted longest, however, when the beans had received only medium roasting. That is bad news for people who like their crema both plentiful and durable; the amount of roasting, which increases one property, diminishes the other. In that sense, espresso is forever a compromise; its beans can be processed to meet one preference or the other but not both at the same time.

Edible foams illustrate a kind of limit to fundamental science. Take milk foam, which depends on the behavior of proteins: chemists have identified the exact proteins in milk, and know how they link up and change under heat. Those microscopic properties explain how the proteins can maintain a foam. But when it comes to making a cappuccino in the kitchen, nothing so profound comes into play; rather, the right or wrong brand of milk can be the determining factor, because different brands may undergo different methods of processing at the dairy.

That suggests a rhetorical question: "Can everyday foams, such as milk foam, ever be fully understood and controlled?" The answer seems to be, fully understood, yes; fully controlled, perhaps not. And for edible foams for which the scientific understanding is less advanced than it is with milk, or which, as in espresso, include a wickedly complex combination of compounds, science cannot even begin to claim full understanding.

By the same token, however, empirical kitchen knowledge, hardly at odds with fundamental chemistry but not much informed by it, is a valuable, useful kind of knowledge. Many scientific characteristics of foam enter into the food and drink that it animates. But when you come down to it, it is the sheer fact of mixing a gas with a liquid or a solid that makes edible foam so pleasing. The mixture provides an airy elegance of presentation that enhances the enjoyment of dining. More important, the mixture gives rise to the unique texture that makes a taste of foamy food or drink a satisfying sensual experience, the elusive quality that food chemists and gourmet tasters call "mouthfeel." How fortunate that "mere" empirical knowledge of such qualities is enough to give us wonderful soufflés, beer, and champagne.

Although it would be a stretch to claim that kitchen knowledge carries over directly into the uses of non-edible foams—from the soapy foams of shaving cream to the quantum foams of spacetime itself—the formation of the latter do depend on similar processes and qualities. The processing that turns a semiliquid mixture of flour and water into solid bread is similar to the procedures that turn a liquid solution of certain molecules into a solid plastic foam. The insulating properties of meringue, which make it possible to create the wonderful contrasts of a baked Alaska, are inherent in the nature of foam; they are the reason you can hold a hot cappuccino in your hand within a plastic foam coffee cup. It is not altogether froth to say that we live in a world of foam, and that contemplating the mysteries of the crema atop a cappuccino may help unlock the secrets of the universe.

Everything Worth Knowing About. . .Ice

No-tech, low-tech, and high-tech

Americans love ice. We make desserts out of it, put it in drinks, and expect motels to provide it by the bucketful. But before ice-making technology, it was a precious commodity laboriously harvested from nature. In the early 19th century, American entrepreneur Frederic Tudor began shipping ice cut from New England ponds to warm climates like the West Indies. He eventually became the "Ice King," selling thousands of tons of ice throughout the United States for iceboxes, which preserved food.

Natural ice remained profitable until a century ago with the arrival of electric refrigerators, which exploited the cooling properties of evaporating liquids. Those white boxes with clumsy cooling units have evolved into sleek stainless steel models, but we still harvest ice. High-end drinkers want premium, artisanal ice in their craft cocktails—perfectly clear, oversized cubes that look good and supposedly melt slowly to avoid dilution. While these can be created in-house, it's a time-consuming process, so specialty companies now provide this elaborate ice. The cubes are hand-cut from big blocks of filtered water gradually frozen in special units for days, a process that eliminates the air bubbles that produce cloudiness. The cost? A whopping $1 each. The Ice King would be green with envy.

A most complicated mineral

Ice is not just the solid form of water. By definition, natural ice is a mineral, like quartz: a naturally occurring inorganic solid with an ordered atomic structure. Or rather, structures. It appears in at least 18 different crystalline forms, plus amorphous forms where the atoms are randomly arranged. The variety that cools our drinks, hexagonal ice or I_h, is the most common type, but it's also quite unusual. Unlike most solids, its frozen state is less dense than its liquid state, thanks to the hexagonal atomic geometry. That means frozen water floats on liquid water—not the case for almost any other substance. The geometry is the reason icebergs float, and helps explain why bodies of water don't freeze solid from the bottom up, allowing aquatic life to survive winters.

Kurt Vonnegut's novel *Cat's Cradle* featured Ice-Nine, a fictional version of ice that would instantly crystallize all water, potentially dooming the world. While there really is an "ice-nine," it forms only under high pressure and low temperature, and does not "lock up" all water.

Common ice also appears as snowflakes, which take on hexagonal forms in atmospheric water vapor. Less well known is hair ice, which grows in fine filaments on wood infested with a specific fungus.

Other exotic ice structures are made in the lab. Researchers recently found that squeezing room-temperature water between atom-thick sheets of carbon locks the water molecules into a two-dimensional square array to form a new kind of ice.

Reshaping the world

Earth's massive ice stores are melting, thanks in part to global warming, and we're seeing the effects on our planet. One of the biggest concerns is the accompanying rise in average global sea level—estimated at 3 to 6 feet by 2100. But other problems pop up with some 400 billion tons of Greenland and Antarctic ice turning into water each year since 2011.

First, there's something called post-glacial rebound. Ice piled onto a landmass presses it down into the squishy mantle layer below Earth's crust. As our world warms, the formerly icy areas rebound, literally raising the land up. That's one reason for the varying sea-level rises along different coasts.

Also, the weight of all that ice had slightly flattened the whole planet. (A single glacier can weigh millions of tons.) Rising temperatures caused it to start becoming spherical again. This influenced Earth's rotation, akin to an ice skater jutting out her arms to spin slower, contributing toward a lengthening of the day by milliseconds over the past 2,500 years. It also alters the planet's axial tilt: The North Pole had been steadily wandering toward Canada, but researchers recently realized that the current ice melt has jolted it on a new course toward the United Kingdom, at an accelerated rate of 7 inches a year since 2000. Astronomers' stellar observations and your phone's GPS program have to take this effect into account.

Water, water everywhere

In 2015, NASA's New Horizons mission to Pluto spectacularly confirmed that the entire solar system boasts water ice, as well as other frozen volatile materials. The composition of these ices sheds light on the properties of their home worlds and gives clues to how the solar system formed some 4.6 billion years ago.

- Near the sun, Mercury has water ice preserved within craters in permanent shadow. (Similar ice appears on Earth's moon.)
- Venus, the hottest planet, can't claim water ice, but astronomers were surprised to find several types of snow were possible.
- Mars' cold temperatures support water ice, including on its North Pole.
- Larger bodies in the asteroid belt, like Ceres and 24 Themis, appear to be full of water ice—in fact, Ceres may have more fresh water than Earth!
- Europa, Enceladus, and other moons of Jupiter and Saturn have ice in abundance. Plates of ice churn and break apart on Europa, while Enceladus boasts an icy, fractured surface.
- Saturn's rings are practically nothing but water ice.
- The ice giants, Uranus and Neptune, likely have rocky cores surrounded by water, methane and ammonia ices.
- Pluto and other Kuiper Belt objects also boast surface and subsurface collections of ice.

Ice could one day power space exploration. Methane, found throughout the solar system, combines with liquid oxygen to yield a type of rocket fuel. That means spacecraft might someday hop through the solar system, refueling from methane ice deposits along the way.

Water's envoy

Clearly, water ice is more abundant than we once thought. But did our solar system just get lucky? And where did it come from in the first place? The answers lie within water itself.

The components of water ice—hydrogen and oxygen atoms—have been around for much of the universe's history, but of course it's not water till they're combined. Astrophysicists think that happened during the earliest days of our solar system, when the sun and planets were nothing more than a swirling cloud of hydrogen and dust particles. If high-energy particles from deep space, called cosmic rays, happened to hit one of those hydrogen atoms, it became ionized, stripped of its electron. Ionized hydrogen atoms could then easily combine with oxygen, forming our H_2O ice.

Planetary scientists tested this by examining the hydrogen atoms in today's water. Deuterium, a type of hydrogen with a neutron, is heavier than regular hydrogen. Scientists can classify water samples, and compare their histories, by learning how many of those hydrogen atoms are deuterium. They've found that the ice on comets has deuterium concentrations very similar to the water in our oceans. The match means our water ice stores have the same cosmic origin as the solar system's comets—among its oldest known objects. So, we know our watery bounty is not unique, since the same conditions likely occurred during the formation of any other solar system. Water is likely similarly abundant around other planets, raising the odds of finding life as we know it, or at least habitable conditions, somewhere else.

The sun has a much lower deuterium concentration than our oceans, suggesting that much of Earth's water comes directly from the initial pre-solar cloud. Just think: Part of every sip of water you drink could be older than the sun.

Ice, ice, baby?

Ice may be responsible for life itself. By analyzing the light from molecular clouds, astronomers have observed not just H_2O, but 200 different molecules including H_2, carbon dioxide, and ammonia—existing either as gases or in ice that coats dust grains. These molecules can interact to produce complex organic compounds, which could generate DNA, RNA, and amino acids. Such biomolecules or their precursors could have ridden to Earth on comets or asteroids to seed life on our planet—and possibly elsewhere.

Scientists at the NASA's Ames Research Center went even further in 2015 when they exposed the organic molecule pyrimidine, found in meteorites, to interstellar conditions. Frozen in ice under ultraviolet radiation, high vacuum, and low temperature, the pyrimidine turned into uracil, cytosine and thymine, major components of earthly DNA and RNA.

While we have not yet found complete biomolecules or their forerunners in molecular clouds, these are promising results. We may soon find that life's true cradle may be the cold and seemingly hostile environment of interstellar ice.

Chapter 2

Technology

Introduction

Technology, the application of scientific principles or other means for practical ends, has a long history in human culture. Today "technology" or "tech" is usually associated with the digital wizardry of the computer and the smart phone, but over the centuries its appearances have been as varied as Neolithic stone tools, Roman aqueducts, steam engines, magnetic resonance imaging (MRI), and genetic engineering. It draws on biomedical as well as physical science, and through its effects on humanity, also involves the social sciences, the humanities, and the arts—all as reflected in the category "Technology."

Lasers and space travel

When the laser was invented in 1960, it seemed a science-fiction "death ray," but has since become a valuable tool with wide applications. Written to celebrate its 50th anniversary, "From Ray-gun to Blu-ray" (2010) covers the history, operation, and uses of the laser. The award of the 2018 Nobel Prize in physics for laser research shows the continuing importance of this device.

Real Scientists Don't Wear Ties: When Science Meets Culture
Sidney Perkowitz
Copyright © 2020 Sidney Perkowitz
ISBN 978-981-4800-68-6 (Hardcover), 978-0-429-35145-7 (eBook)
www.jennystanford.com

Space technology has reached a level that allows us to explore the solar system out to Pluto and beyond, and "How Close Are We to Actually Becoming Martians?" (2015) (written for the release of the popular film *The Martian*) shows that we may put people on Mars in just a few years. But the next step, leaving our solar system to reach stars other than our sun is currently unattainable and may remain so, as "Ad Astra! To the Stars!" (2012) discusses. Yet the ancient desire to reach the stars only grows as we continue to find more exoplanets orbiting distant suns.

Technology in the clinic

Medical technology draws on physical and biomedical science, for instance in scanning technologies such as MRI. "Brain Injuries in Soccer" (2016) shows how physical analysis and MRI help us understand the serious brain trauma due to concussion that first came to wide notice in American professional football and also shows up in soccer. "When Vision Betrays" (2017) relates how optics and materials science, combined with precision surgery, led to the routine replacement of cataract-ridden natural eye lenses with artificial plastic versions. Today new developments in materials, and in laser and ultrasound technology, enhance the procedure.

Robots and artificial intelligence

Rapid development in robotics and artificial intelligence (AI) is changing the world. In "John Markoff's Love for 'Machines'" (2015) and "Removing Humans from the AI Loop: Should We Panic?" (2016), I comment on two books by a technology writer and a roboticist respectively. These books survey the recent history of the technology and its uses such as self-driving cars; how robots and AI affect society as they replace human workers; and the possibility of creating powerful AIs with their potential dangers. "Do We Have Moral Obligations to Robots?" focuses instead on the ethical questions that the creation of truly life-like and perhaps self-conscious beings would raise.

Technology, society, and human behavior

Technology changes us and society. "The Internet Before the Internet: Paul Otlet's Mundaneum" (2016) describes early schemes to store and distribute human knowledge until the internet finally made that widely possible. "The Internet of Things: Totally New and a Hundred Years Old" (2015) draws on the remarkable science-fiction story "The Machine Stops" (1909) by the great English novelist E. M. Forster to consider our growing physical dependence on digital technology. "Crimes of the Future" (2016) notes the rising use of computer algorithms to pre-identify criminals and predict crime much as in the Tom Cruise film *Minority Report* (2002), with troubling implications for civil rights.

"How to Understand the Resurgence of Eugenics" (2017) points out how our growing ability to control human qualities through gene editing—as in the recent efforts by a Chinese scientist to modify human babies—could lead to a new Nazi-like eugenics. "The Case Against an Autonomous Military" (2018) points to another frightening possibility, the development of AI-driven battlefield weapons that choose their own targets.

"Frankenstein Turns 200 and Becomes Required Reading for Scientists" (2018) uses the 200th anniversary of the Frankenstein story to comment on the ethical quandaries that science and technology face today. Finally, "Can a Physics of Panic Explain the Motions of the Crowd?" (2018) shows how physical analysis and data analysis can predict human behavior.

Future technologies

I speculate on possibilities for three appealing but imaginary technologies that we wish would become real—invisibility, teleportation, and tractor beams. In 2011 I wrote about them in a series called "Fantasy into Science" for the National Academy of Sciences Science and Entertainment Exchange, extending existing real science as far as I reasonably could toward the science-fiction dreams. We can now realize all three fantasies on small and limited scales, but we are not yet making Harry Potter invisibility cloaks, teleporting large objects, or creating powerful force beams. With ongoing research in all these areas, though, we can always hope for more.

From Ray-gun to Blu-ray

There is one particular scene in H. G. Wells' 1898 tale *The War of the Worlds* that, if only I had remembered it, could have helped me to avoid a bad moment in my laser lab in 1980. In the story—published long before lasers came along in 1960—the Martians wreak destruction on earthlings with a ray that the protagonist calls an "invisible, inevitable sword of heat," projected as if an "intensely heated finger were drawn...between me and the Martians." In all but name, Wells was describing an infrared laser emitting an invisible straight-line beam—the same type of laser that, decades later in my lab, burned through a favorite shirt and started on my arm.

Wells' bold prediction of a destructive beam weapon preceded many others in science fiction. From the 1920s and 1930s, Buck Rogers and Flash Gordon wielded eye-catching art-deco ray-guns in their space adventures as shown in comics and in films. In 1951 the powerful robot Gort projected a ray that neatly disposed of threatening weapons in the film *The Day the Earth Stood Still*. Such appearances established laser-like devices in the popular mind even before they were invented. But by the time the evil Empire in *Star Wars Episode IV: A New Hope* (1977) used its Death Star laser to destroy an entire planet, lasers were a thing of fact, not just fiction. Lasers were changing how we live, sometimes in ways so dramatic that one might ask, which is the truth and which the fiction?

Like the fictional science, the real physics behind lasers has its own long history. One essential starting point is 1917, when Einstein, following his brilliant successes with relativity and the theory of the photon, established the idea of stimulated emission, in which a photon induces an excited atom to emit an identical photon. Almost four decades later, in the 1950s, the U. S. physicist Charles Townes used this phenomenon to produce powerful microwaves from a molecular medium held in a cavity. He summarized the basic process—microwave amplification by stimulated emission of radiation—in the acronym "maser."

After Townes and his colleague Arthur Schawlow proposed a similar scheme for visible light, Theodore Maiman, of the Hughes Research Laboratories in California, made it work. In 1960 he amplified red light within a solid ruby rod to make the first laser.

Its name was coined by Gordon Gould, a graduate student working at Columbia University, who took the word "maser" and replaced "microwave" with "light," and later received patent rights for his own contributions to laser science.

Following Maiman's demonstration of the first laser there was much excitement and enthusiasm in the field, and the ruby laser was soon followed by the helium neon or HeNe laser, invented at Bell Laboratories in 1960. Capable of operating as a small, low-power unit, it produced a steady, bright-red emission at 633 nm. However, an even handier type was discovered two years later when a research group at General Electric saw laser action from an electrical diode made of the semiconductor gallium arsenide. That first laser diode has since mushroomed into a versatile family of small devices that covers a wide range of wavelengths and powers. The diode laser quickly became the most prevalent type of laser, and still is to this day—according to a recent market survey, 733 million of them were sold in 2004.

Better living through lasers

As various types of lasers became available, and different uses for them were developed, these devices entered our lives to an extraordinary extent. While Maiman was dismayed that his invention was immediately called a "death ray" in a sensationalist newspaper headline, lasers powerful enough to be used as weapons would not be seen for another 20 years. Indeed, the most widespread versions are compact units typically producing mere milliwatts.

A decade and a half after their invention, HeNe lasers, and then diode lasers, would become the basis of bar-code scanning—the computerized registration of the black and white pattern that identifies a product according to its universal product code (UPC). The idea of automating such data for use in sales and inventory originated in the 1930s, but it was not until 1974 that the first in-service laser scanning of an item with a UPC symbol—a pack of Wrigley's chewing gum—occurred at a supermarket checkout counter in Ohio. Now used globally in dozens of industries, bar codes are scanned billions of times daily and are claimed to save billions of dollars a year for consumers, retailers, and manufacturers alike.

The diverse uses of lasers

Lasers would also come to dominate the way in which we communicate. They now connect many millions of computers around the world by flashing binary bits into networks of pure-glass optical fiber at rates of terabytes per second. Telephone companies began installing optical-fiber infrastructure in the late 1970s and the first transatlantic fiber-optic cable began operating between the United States and Europe in 1988, with tens of thousands of kilometers of undersea fiber-optic cabling now in existence worldwide. This global web is activated by laser diodes, which deliver light into fibers with core diameters of a few micrometers at wavelengths that are barely attenuated over long distances. In this role, lasers have become integral to our interconnected world.

As lasers grew in importance, their fictional versions kept pace with—and even enhanced—the reality. Only four years after the laser was invented, the film *Goldfinger* (1964) featured a memorable scene that had every man in the audience squirming: Sean Connery as James Bond is tied to a solid gold table along which a laser beam moves, vaporizing the gold in its path and heading inexorably toward Bond's crotch—though as usual, Bond emerges unscathed.

That laser projected red light to add visual drama, but its ability to cut metal foretold the invisible infrared beam of the powerful carbon-dioxide (CO_2) laser—the type that once ruined my shirt. Invented in 1964, CO_2 lasers emitting hundreds of watts in continuous operation were introduced as industrial cutting tools in the 1970s. Now, kilowatt versions are available for uses such as "remote welding" in the automobile industry, where a laser beam directed by steerable optics can rapidly complete multiple metal spot welds. High-power lasers are suitable for other varied industrial tasks, and even for shelling nuts.

Digital media

Aside from the helpful and practical uses of lasers, what have they done to entertain us? For one thing, lasers can precisely control light waves, allowing sound waves to be recorded as tiny markings in digital format and the sound to be played back with great fidelity.

In the late 1970s, Sony and Philips began developing music digitally encoded on shiny plastic "compact discs" (CDs) 12 centimeters in diameter. The digital bits were represented by micrometer-sized pits etched into the plastic and scanned for playback by a laser diode in a CD player. In retrospect, this new technology deserved to be launched with its own musical fanfare, but the first CD released, in 1982, was the commercial album *52nd Street* by rock artist Billy Joel.

In the mid-1990s the CD's capacity of 74 minutes of music was greatly extended via digital versatile discs or digital video discs (DVDs) that can hold an entire feature-length film. In 2009 Blu-ray discs (BDs) appeared as a new standard that can hold up to 50 gigabytes, which is sufficient to store a film at exceptionally high resolution. The difference between these formats is the laser wavelengths used to write and read them—780 nm for CDs, 650 nm for DVDs, and 405 nm for BDs. The shorter wavelengths give smaller diffraction-limited laser spots, which allow more data to be fitted into a given space.

Although the download revolution has led to a decline in CD sales—27% of music revenue last year was from digital downloads—lasers remain essential to our entertainment. They carry music, films and everything that streams over or can be downloaded via the internet and telecoms channels, depositing them into our computers, smart phones, and other digital devices.

Death rays. . .

Among the films that you might choose to download over the internet are some in which lasers are portrayed as destructive devices, encouraging negative connotations. In the film *Real Genius* (1985), a scientist co-opts two brilliant young students to develop an airborne laser assassination weapon for the military and the CIA. The students avenge themselves by sabotaging the laser to heat a huge vat of popcorn, producing a tsunami of popped kernels that bursts open the scientist's house. The film *RoboCop* (1987) shows a news report that a malfunctioning U. S. laser in orbit around the Earth has wiped out part of southern California. This was a satirical response to the idea of laser weapons in space, a hotly pursued dream for then U. S. President Ronald Reagan.

The U. S. military was thinking about laser weapons well before high-power industrial CO_2 lasers were melting metal. As the Cold War raised fears of all-out conflict with the Soviet Union, the potential for a new hi-tech weapon stimulated the Pentagon to fund laser research even before Maiman's result. But it was difficult to generate enough beam power within a reasonably sized device— early CO_2 lasers with kilowatt outputs were too unwieldy for the battlefield. Eventually, in 1980, the Mid-Infrared Advanced Chemical Laser reached pulsed powers of megawatts, but was still a massive device. Even worse, absorption and other atmospheric effects made its beam ineffective by the time it reached its target.

That would not be a concern, however, for lasers fired in space to destroy nuclear-tipped intercontinental ballistic missiles (ICBMs) before they re-entered the atmosphere. Development of suitably powerful lasers such as those emitting X-rays became part of the multibillion-dollar anti-ICBM Strategic Defense Initiative (SDI) proposed by Reagan in 1983. Known to the general public and even to scientists and the government as "Star Wars" after the film, the scheme had an undeniably science-fiction flavor. But the U. S. weaponization of space was never realized—by the 1990s technical difficulties and the fall of the Soviet Union had turned laser-weapons development elsewhere. Now it is mostly directed toward smaller weapons such as airborne lasers that have a range of hundreds of kilometers.

. . .and life rays

While the morality associated with weapons may be debatable, lasers are used in many other areas that are undeniably good, such as medicine. The first medical use of a laser was in 1961, when doctors at Columbia University Medical Center in New York destroyed a tumor on a patient's retina with a ruby laser. Because a laser beam can enter the eye without injury, ophthalmology has benefited in particular from laser methods, but their versatility has also led to laser diagnosis and treatment in other medical areas.

Using CO_2 and other types of lasers with varied wavelengths, power levels, and pulse rates, doctors can precisely vaporize tissue, and can also cut tissue while simultaneously cauterizing it to reduce surgical trauma. One example of medical use is LASIK (laser-assisted *in situ* keratomileusis) surgery in which a laser beam reshapes the cornea to correct faulty vision. By 2007, some 17 million people worldwide had undergone the procedure.

In dermatology, lasers are routinely used to treat benign and malignant skin tumors, and also to provide cosmetic improvements such as removing birthmarks or unwanted tattoos. Other medical uses are as diverse as treating inaccessible brain tumors with laser light guided by a fiber-optic cable, reconstructing damaged or obstructed fallopian tubes and treating herniated discs to relieve lower-back pain, a procedure carried out on 500,000 patients per year in the United States.

Yet another noble aim of using lasers is in basic and applied research. One notable example is the National Ignition Facility (NIF) at the Lawrence Livermore National Laboratory in California. NIF's 192 ultraviolet laser beams, housed in a stadium-sized, 10-story building, are designed to deliver a brief laser pulse measured in hundreds of terawatts into a millimeter-scale, deuterium-filled pellet. This is expected to create conditions like those inside a star or a nuclear explosion, allowing the study of both astrophysical processes and nuclear weapons.

A more widely publicized goal is to induce the hydrogen nuclei to fuse into helium, as happens inside the Sun, to produce an enormous energy output. After some 60 years of effort using varied approaches, scientists have yet to achieve fusion power that produces more energy than a power plant would need to operate. If laser fusion were to successfully provide this limitless, non-polluting energy source, that would more than justify the overruns that have brought the cost of NIF to $3.5 billion. Although some critics consider laser fusion a long shot, recent work at NIF has realized some of its initial steps, increasing the odds for successful fusion.

Popular culture is also hopeful about the role of lasers in "green" power. Although the film *Chain Reaction* (1996) badly scrambles the science, it does show a laser releasing vast amounts of clean energy from the hydrogen in water. In *Spider-Man 2* (2004), physicist Dr. Octavius uses lasers to initiate hydrogen fusion that will supposedly help humanity; unfortunately, this is no advertisement for the benefits of fusion power, for the reaction runs wild and destroys his lab.

Lasers in high and not-so-high culture

Situated between the ultra-powerful lasers meant to excite fusion and the low-power units at checkout counters are lasers with mid-range powers that can provide highly visible applications in art and entertainment, as artists quickly realized. A major exhibit of laser art was held at the Cincinnati Museum of Art as early as 1969, and in 1971 a sculpture made of laser beams was part of the noted "Art and Technology" show at the Los Angeles County Museum of Art. In 1970 the well-known U. S. artist Bruce Nauman presented "Making Faces," a series of laser hologram self-portraits, at New York City's Finch College Museum of Art.

Other artists followed suit in galleries and museums, but lasers have been most evident in larger venues. Beginning in the late 1960s, beam-scanning systems were invented that allowed laser beams to dynamically follow music and trace intricate patterns in space. This led to spectacular shows such as that at the Expo '70 World's Fair in Osaka, Japan, and those in planetariums. A favorite type featured "space" music, like that from *Star Wars*, accompanied by laser effects.

Rock concerts by Pink Floyd and other groups were also known for their laser shows, though these are now tightly regulated because of safety issues. But spectacular works of laser art continue to be mounted, for example the outdoor installations "Photon 999" (2001) and "Quantum Field X3" (2004) created at the Guggenheim Museum in Bilbao, Spain, by Japanese-born artist Hiro Yamagata, and the collaborative Hope Street Project, installed in 2008. This linked together two major cathedrals in Liverpool, U. K., by intense laser beams—one highly visible green beam and also several invisible ones—that carried voices and generated ambient music to be heard at both sites.

After 50 years, striking laser displays can still evoke awe, and lasers still carry a science-fiction-ish aura, as demonstrated by hobbyists who fashion mock ray-guns from blue laser diodes. Unfortunately, the mystique also attaches itself to products such as the so-called quantum healing cold laser, whose grandiose title uses scientific jargon to impress would-be customers. Its maker, Scalar Wave Lasers, asserts that its 16 red and infrared laser diodes provide substantial health and rejuvenation benefits. Even the word "laser" has been appropriated to suggest speed or power, such as for the popular Laser Class small sailboats and the Chrysler and Plymouth Laser sports cars sold from the mid-1980s to the early 1990s.

The laser's distinctive properties have also become enshrined in language. A search of the massive Lexis Nexis Academic research database (which encompasses thousands of newspapers, wire services, broadcast transcripts, and other sources) covering the last two years yields nearly 400 references to phrases such as "laser-like focus" (appearing often enough to be a cliché), "laser-like precision," "laser-like clarity," and, in a description of Russian Prime Minister Vladimir Putin expressing his displeasure with a particular businessman, "laser-like stare."

Lasers have significantly influenced both daily life and science. With masers, they have been part of research, including work outside laser science itself, that has contributed to more than 10 Nobel Prizes, beginning with the 1964 physics prize awarded to Charles Townes with Aleksandr Prokhorov and Nicolay Basov for their fundamental work on lasers. Other related Nobel-Prize research includes the invention of holography and the creation of the first Bose–Einstein condensate, which was made by laser cooling a cloud of atoms to ultra-low temperatures. Also, in dozens of applications from Raman spectroscopy to adaptive optics for astronomical telescopes, lasers continually contribute to how science is done. They are also essential for research in such emerging fields as quantum entanglement and slow light.

It is a tribute to the scientific imagination of the laser pioneers, as well as to the literary imagination of writers such as H. G. Wells, that an old science-fiction idea has come so fully to life. But not even imaginative writers foresaw that Maiman's invention would change the music business, create glowing art, and operate in supermarkets across the globe. In the cultural impact of the laser, at least, truth really does outdo fiction.

How Close Are We to Actually Becoming Martians?

Like any long-distance relationship, our love affair with Mars has had its ups and downs. The planet's red tint made it a distinctive—but ominous—nighttime presence to the ancients, who gazed at it with the naked eye. Later we got closer views through telescopes, but the planet still remained a mystery, ripe for speculation.

A century ago, the American astronomer Percival Lowell mistakenly interpreted Martian surface features as canals that intelligent beings had built to distribute water across a dry world. This was just one example in a long history of imagining life on Mars, from H. G. Wells portraying Martians as bloodthirsty invaders of Earth, to Edgar Rice Burroughs, Kim Stanley Robinson, and others wondering how we could visit Mars and meet the Martians.

One recent entry in this long tradition is the sci-fi flick *The Martian*, released on October 2, 2015. Directed by Ridley Scott and based on Andy Weir's self-published novel, it tells the story of an astronaut (played by Matt Damon) stranded on Mars. Both book and movie try to be as true to the science as possible—and, in fact, the science and the fiction around missions to Mars are rapidly converging.

NASA's Curiosity rover and other instruments have shown that Mars once had oceans of liquid water, a tantalizing hint that life was once present. And now NASA has just reported the electrifying news that liquid water is flowing on Mars today.

This discovery increases the odds that there is currently life on Mars—picture microbes, not little green men—while heightening interest in NASA's proposal to send astronauts there by the 2030s as the next great exploration of space and alien life.

So how close are we to actually sending people to Mars and having them survive on an inhospitable planet?

First, we have to get there

Making it to Mars won't be easy. It's the next planet out from the sun, but a daunting 140 million miles away from us, on average—far beyond the Earth's moon, which, at nearly 250,000 miles away, is the only other celestial body human beings have set foot on.

Nevertheless, NASA and several private ventures believe that by further developing existing propulsion methods, they can send a manned spacecraft to Mars.

One NASA scenario would, over several years, pre-position supplies on the Martian moon Phobos, shipped there by unmanned spacecraft; land four astronauts on Phobos after an eight-month trip from Earth; and ferry them and their supplies down to Mars for a 10-month stay, before returning the astronauts to Earth.

We know less, though, about how a long voyage inside a cramped metal box would affect crew health and morale. Extended time in space under essentially zero gravity has adverse effects, including loss of bone density and muscle strength, which astronauts experienced after months aboard the International Space Station (ISS).

There are psychological factors, too. ISS astronauts in Earth orbit can see and communicate with their home planet, and could reach it in an escape craft, if necessary. For the isolated Mars team, home would be a distant dot in the sky; contact would be made difficult by the long time lag for radio signals. Even at the closest approach of Mars to the Earth, 36 million miles, nearly seven minutes would go by before anything said over a radio link could receive a response.

To cope with all this, the crew would have to be carefully screened and trained. NASA is now simulating the psychological and physiological effects of such a journey in an experiment that is isolating six people for a year within a small structure in Hawaii.

Surviving in an inhospitable Martian landscape

These concerns would continue during the astronauts' stay on Mars, which is a harsh world. With temperatures that average –80 Fahrenheit (–62 Celsius) and can drop to –100 Fahrenheit (–73 Celsius) at night, it is cold beyond anything we encounter on Earth; its thin atmosphere, mostly carbon dioxide (CO_2), is unbreathable and supports huge dust storms; it is subject to ultraviolet radiation from the sun that may be harmful; and its size and mass give it a gravitational pull that is only 38% of the Earth's—which astronauts exploring the surface in heavy protective suits would welcome, but could also further exacerbate bone and muscle problems.

As the astronauts establish their base, NASA is planning to use Mars' own resources to overcome some of these obstacles.

Fortunately, water and oxygen should be available. NASA had planned to try a form of mining to retrieve water existing just below the Martian surface, but the new finding of surface water may provide an easier solution for the astronauts. Mars also has considerable oxygen bound up in its atmospheric CO_2. In the MOXIE process (Mars OXygen In situ resource utilization Experiment), electricity breaks up CO_2 molecules into carbon monoxide and breathable oxygen. NASA proposes to test this oxygen factory aboard a new Mars rover in 2020 and then scale it up for the manned mission.

There is also potential to produce the compound methane from Martian sources as rocket fuel for the return to Earth. The astronauts should be able to grow food, too, using techniques that recently allowed the ISS astronauts to taste the first lettuce grown in space.

Without utilizing some of Mars' raw materials, NASA would have to ship every scrap of what the astronauts would need: equipment, their habitation, food, water, oxygen, and rocket fuel for the return trip. Every extra pound that has to be hauled up from Earth makes the project that much more difficult. "Living off the land" on Mars, though it might affect the local environment, would hugely improve the odds for success of the initial mission—and for eventual settlements there.

NASA will continue to learn about Mars and hone its planning over the next 15 years. Of course, there are formidable difficulties ahead; but it's key that the effort does not require any major scientific breakthroughs, which, by their nature, are unpredictable. Instead, all the necessary elements depend on known science being applied via enhanced technology.

Yes, we're closer to Mars than many may think. And a successful manned mission could be the signature human achievement of our century.

Ad Astra! To the Stars!

An alien spacecraft scouting out Earth's scientific prowess in late September 2011 may well have zeroed in on NASA's Kennedy Space Center in Florida. But the aliens might have learned more if they had flown some miles west to the 100 Year Starship Study (100YSS) conference in Orlando. There they would have seen that human space technology is limited, but in observing the event's hundreds of attendees—from ex-astronauts and engineers to artists, students, and science-fiction writers—the aliens would also have encountered humanity's adventurous, stubborn, mad, and glorious aspiration to reach the stars.

Maybe this desire to literally travel to the stars by spaceship arises because these distant suns have always seemed to offer a high and remote plane of existence. Aristotle in fact placed the fixed stars furthest from Earth—the center of his cosmology—and nearest the Prime Mover that causes cosmic motion. The phrase *"sic itur ad astra,"* or "thus one goes to the stars"—from the Roman poet Virgil—refers to reaching divinity or immortality. But another phrase—*"per aspera ad astra"* or "through hardships to the stars"—reminds us that they are not easy to reach, except in science fiction that sidesteps the difficulties caused by the vast distances the journeys would entail.

Now, with the exploration of the solar system by the U. S. space agency NASA and others well under way, and with the discovery of hundreds of exoplanets orbiting distant stars, it may be time to contemplate the next great jump outwards.

"To boldly go": but not yet

100YSS was the first conference to enable experts, enthusiasts and the general public to gather and seriously consider interstellar travel. Surprisingly, it was not sponsored by NASA but by the Defense Advanced Research Projects Agency (DARPA) of the U. S. Department of Defense, which is also putting money into the effort. DARPA supports novel military science, and its willingness to look at seemingly "fringe" ideas has paid off in the past, although building a starship might seem beyond even its wide embrace.

But as pointed out at the meeting by DARPA's David Neyland, who started and organized 100YSS, military and civilian applications have come from advances in robotics, materials, and other areas developed for use in space. Having sponsored other space-related work as well, DARPA has faith that unimaginable new ideas will emerge from a project to design and perhaps build a starship within a century. Although it does not necessarily envisage reaching a star in that time, the project would have to draw upon the very best in science and society.

It is a cosmic irony that although we thought sending people and machines through the solar system was the hard part, it was actually the easy part, compared with what it will take to cross the huge void between us and the stars. Current propulsion technology moves a spacecraft at only 0.005% of the speed of light, or $0.00005c$. That means a trip of some 80,000 years even to Alpha Centauri, the star system nearest the Sun but hardly a close neighbor at more than four light-years' distance. For comparison, the furthest traveling human-made object ever, NASA's Voyager 1 spacecraft, has in the 34 years since its launch in 1977 penetrated just 0.002 light-years into space.

That speed of $0.00005c$ can probably be improved but only to values still well below c, so a spacecraft aimed at near or distant stars would have to maintain its inhabitants for decades or millennia. Launching or even seriously developing such a miniature world would require a massive investment in research, and in material and human resources. But before we get caught up in these details we must first figure out and overcome the problem of propulsion.

Getting up to speed

A starship needs a rocket engine that efficiently develops thrust, because the craft must accelerate for a long time to reach high velocity. This runs headlong into a catch-22: long-term acceleration means a craft crammed with fuel, the mass of which resists acceleration and allows only a small payload. Chemical rockets such as the Saturn—which carried humanity to the Moon in three stages and burned a kerosene derivative or liquid hydrogen with liquid oxygen—just will not do. These produce high thrust but need lots of fuel to do so, giving a small push per kilogram of fuel.

This reasoning is quantified in a famous version of Newton's third law, derived in 1903 by Russian rocket pioneer Konstantin Tsiolkovsky, which relates a rocket's speed to its thrust and fuel load. Using appropriate parameters, the rocket equation delivers the *coup de grâce*: like a camel that cannot carry enough feed to nourish itself as it plods into the desert, a chemical rocket cannot possibly carry enough fuel to keep going and reach a respectable fraction of c.

To reduce the fuel load, rocketeers are therefore exploring more efficient fuels as measured by the energy per kilogram they supply. The most effective source is matter–antimatter annihilation, which sounds like science fiction and in fact does power the spacecraft in *Star Trek*. The advantage of this process is that it yields the maximum possible energy-to-mass ratio of c^2 as it fully converts one into the other according to $E = mc^2$. But since we have to date made only fractions of nanograms of antimatter, in CERN's gigantic Large Hadron Collider, antimatter propulsion does not seem to be a real possibility.

Nuclear fuels are less efficient than matter–antimatter annihilation but still yield millions of times the energy density of chemical fuels. In nuclear-pulse propulsion, proposed in the 1940s, a starship would drop fission or fusion bombs behind itself and detonate them against a pusher plate to thrust itself forward. Indeed, in 1958, under DARPA's Project Orion, Freeman Dyson of the Institute for Advanced Study designed a massive craft driven to $0.033c$ by hundreds of thousands of thermonuclear bombs. It could supposedly have been built with technology then current, but fortunately the 1963 Nuclear Test Ban treaty prevented further development of this frightful brute-force method.

Fusion power and lasers

A more refined approach came in 1973 from the British Interplanetary Society (BIS), a private group of "spaceflight enthusiasts" founded in 1933. Its Project Daedalus explored the use of nuclear fusion in a reaction chamber to reach the stars within a human lifetime, though without any humans. Volunteer scientists and engineers designed an unmanned probe to examine Barnard's Star, 5.9 light-years away, which supposedly had an orbiting planet (now known to be nonexistent). The ship's 53,000 tonnes were planned to consist mostly of deuterium and helium-3 fuel, with only 450 tonnes of payload; but at a speed of $0.12c$ it would reach its target in 50 years.

Another approach, which amazingly needs no fuel at all, harks back to Johannes Kepler, who in 1619 correctly surmised that light deflects comets' tails. Photons can push a sail to drive a spacecraft, as demonstrated in 2010 by the Japan Aerospace Exploration Agency (JAXA). After launch, its IKAROS spacecraft unfolded a 200 m^2 sail and was accelerated by sunlight. (IKAROS stands for Interplanetary Kite-craft Accelerated by Radiation of the Sun—a play on Greek mythology's Icarus, son of Daedalus, who flew too near the Sun.) Though the power of the Sun drops off with distance squared, a tight laser or microwave beam could push a sail harder and longer. According to one estimate presented at 100YSS, a laser with terawatts of power could bring a craft to $0.13c$.

These methods are under further study. In 2009 members of BIS and the Tau Zero Foundation, another private group, initiated Project Icarus to update fusion propulsion as proposed in Project Daedalus. But decades of scientific effort have yet to yield fusion that actually produces a net amount of energy, though laser inertial confinement, now being tested at the National Ignition Facility (NIF) at the Lawrence Livermore National Laboratory in California, looks promising. As it happens, NIF also shows that beamed propulsion could be feasible since its lasers are planned to deliver terawatts of power. However, NIF is stadium-sized and cost billions, so a purpose-built terawatt beam source would be a major undertaking.

Faster than light

Even if these methods can be developed fairly soon, enthusiasts with bigger dreams would like to go beyond speeds of around $0.1c$ and even exceed c. But since faster-than-light (FTL) travel violates special relativity, this is where reasonably solid propulsion science becomes speculative or "exotic," as it is tactfully called.

Science fiction has long used exotic methods such as "warp drive," which enables FTL travel in *Star Trek* but in fact originated much earlier. In 1931 John W. Campbell (later to exert major influence as editor of the magazine *Astounding Science Fiction*) used the concept of distorted spacetime from general relativity to introduce FTL travel in his story "Islands of space." Its heroes enclose their spaceship in a warped "hyperspace" that allows it to move astoundingly fast, reaching Alpha Centauri in a mere fifth of a second.

General relativity really does in principle offer ways to evade the speed limit. In 1994 it inspired an FTL approach by theoretical physicist Miguel Alcubierre at the University of Wales that resembles Campbell's method. His idea was to contract spacetime in front of a spaceship and expand spacetime behind it, creating a bubble that propels the craft at any speed without violating special relativity. Although the mathematics is impeccable, this seductive idea requires negative mass, which does not exist as far as we know, let alone in the astronomical quantities needed for an actual drive.

This and other approaches were examined in NASA's Breakthrough Propulsion Physics (BPP) program, which ran from 1996 to 2002 and sought new ways to make interstellar travel feasible. In 2008 the BPP's director Marc Millis concluded that "no breakthroughs appear imminent." Three years later, the Alcubierre drive, cosmic wormholes, quantized inertia and other exotica received the same verdict at 100YSS: James Benford of Microwave Sciences, who chaired and summarized the propulsion sessions, characterized the speculative methods as currently being "a bridge too far." (The same can be said of quantum entanglement, which was presented at 100YSS as a potential means of FTL communication—contrary to current scientific understanding.)

Healthy, happy humans

For the foreseeable future, it looks as though we will be stuck with speeds near $0.1c$ at most, with protracted interstellar travel times. So, to deal with the distinct possibility that starship crews would have to function onboard for decades or more, 100YSS included sessions about alternatives to cramming people into a steel box for long periods, and about building "generation" ships if that proves necessary. Alternatives include suspended animation, and unmanned craft that could report back or carry the DNA and other resources needed to recreate humans on arrival at an exoplanet.

But sending whole functioning people, while keeping them healthy, sane, and motivated in a closed and isolated world (radio traffic with Earth would be delayed by a year each way for every light-year the ship travels) raises lots of issues. Some of these have been foreseen in science fiction, as in Robert Heinlein's cautionary tale *Universe* from 1941. As the book's blurb puts it, "Their world was a giant spaceship, its purpose and destination lost in centuries of drifting among the stars." To make conditions even more dire, the cover shows two male crew members apparently in good shape and with nicely combed hair—except that one of them sports two heads!

Exaggerated though this is, damaging radiation that could produce mutations is just one of the problems to be faced in a long-term artificial environment. Along with propulsion, these would make planning, building, and crewing a long-haul starship the most complex scientific project ever. Sessions at 100YSS considered how to manage such an effort, dealing with questions including how to elicit the best technology and where to find funding. To kick off the project, DARPA favors the private route: it is offering $500,000 to develop a "non-government organization for persistent, long-term, private-sector investment into the myriad of disciplines needed to make long-distance space travel viable."

Should we, or shouldn't we?

Despite all the science at 100YSS, building a starship was more than once compared with constructing a great medieval cathedral over many years. After all, there is a certain religious or spiritual dimension to the fundamental question: why seek the stars?

Indeed, some speakers at 100YSS saw great spiritual benefits to interstellar travel. They included Anousheh Ansari, a businesswoman and the first Iranian in space, who felt transformed after her experience in 2006 as a private space traveler, and Thomas Hoffmann, a protestant pastor from Tulsa, Oklahoma, who saw travel to the stars as carrying religious feeling into a new sacred space. Others spoke of a "moral imperative" to start anew by escaping an industrial civilization that has despoiled our planet, or of a back-up plan in case of global disaster. And always, there is the part of the human spirit that would speed off to explore the universe simply "because it is there."

The romantic quest has a strong pull, but given the obstacles, it is fair to ask whether the interstellar dream is actually a bridge too far. Attendees at 100YSS were true believers, but does the rest of humanity share the dream? To put the project in context, what definite need or benefit would convince private investors or governments to provide vast sums to reach the stars, especially amid economic uncertainty?

Yet, like building a cathedral, building a starship could rally humanity to join together in a common cause. And like the late Steve Jobs, perhaps a true visionary could discern the yearnings of millions and give them what they want before they know they want it—a starship, instead of iPads. But the visionary would also need an accompanying effort that replaces "*ad astra*" with a motto from the U. S. Navy Seabees, the construction battalions known for doing what needs to be done in record time: "The difficult we do at once; the impossible takes a bit longer."

Brain Injuries in Soccer

American football is a rough game where big, powerful players slam their opponents into the ground. As tough athletes who are additionally protected by pads and helmets, players generally jump up from these crashes and pile-ups ready for the next play, though sometimes one hurts a knee or leg and has to be carried off the field. But far worse can happen if hits to a player's head during play cause concussions that shake the brain, leading to hidden but serious long-term cognitive deterioration. Now an even more popular sport, soccer, is facing the same alarming fact that it may be exposing its players to terrible consequences.

Research is underway to make football and soccer safer, and to understand why shaking the brain can harm it so greatly. Evidence that this happens in American football has been building for a long time. But 4,500 former players had to sue the National Football League (NFL), the multi-billion-dollar business that controls professional American football, to make it confront the issue. In 2015 a Federal judge approved a final settlement, in which the NFL agreed to spend $765 million to compensate ex-players with damaged brains and to fund research in brain injuries. The NFL has also changed its rules to better deal with head impacts on the field.

Many observers think this is not enough, calling the NFL data on concussions flawed. Moreover, the problem also affects high school and college students who play American football as amateurs. Still, the issue is beginning to be recognized. Now it's the turn of soccer, which is far more popular than American football with a quarter of a billion male and female players world-wide and hordes of devoted fans, to deal with the same unwelcome reality.

It may seem a stretch to compare soccer to football in terms of potential for brain injury. In football, the amazingly loud "clunk" heard when two helmeted heads collide shows that considerable force and energy are being transferred. Any impact to the head can sufficiently rattle the brain inside it to produce a concussion (the Latin root of the word means "to shake violently") and traumatic brain injury (TBI) with consequences such as unconsciousness and mental confusion as well as long-term harm.

What is more troubling is that even relatively mild traumatic brain injury (mTBI), which a player might "walk off" before returning to the game, may also produce cognitive degeneration if experienced repeatedly. The degenerative condition, chronic traumatic encephalopathy (CTE), resembles Alzheimer's disease with symptoms of depression, memory loss, and emotional instability. Like Alzheimer's, it is untreatable and appears only after years, complicating its diagnosis. An important clue to identifying CTE in football was the discovery of former players of American football whose mental state had degraded, sometimes at an early age.

Compared to football, soccer is the elegant game where speed, agility, and skill count more than brute strength; where players wear no helmets and pads but are lightly clad with only shin guards as armor; and where players can be ejected from a game for excessive force. Yet soccer has plenty of potential for injuries. Its fast-moving players crash into each other, hit the ground hard after tripping or diving for the ball, or run into a goal post. During the 2014 World Cup in Brazil, three players were knocked unconscious by hits to the head from other players, but insisted on returning to play against medical advice. One, Germany's Christoph Kramer, showed mental disorientation characteristic of concussion and had to be helped from the field.

To make matters worse, soccer has a unique feature that seems almost perversely designed to hurt the head. Besides controlling the ball with the feet, a soccer player controls it by heading it, that is, letting it bounce off the head. This subjects the head and brain to forces and twists these body parts were never meant to endure on a regular basis. The effects of these impacts could build up over time to foster CTE, especially for young players and women, whose smaller or weaker necks may not sufficiently support their heads to safely absorb these stresses.

Just how much damage heading causes in soccer is still unresolved. But like football, some particular cases point to real problems with TBI, mTBI, and CTE. In 2013, after two prominent soccer figures—an ex-World Cup Brazilian player and an American semi-professional player aged 29—died, examination of their brains showed that they had suffered from CTE. Similar observations go back further. In 2002, Jeff Astle, a player for the West Bromwich Albions soccer club in the U. K. known for his prowess in heading, died at age 59. His brain was so raddled by CTE that a physician thought it resembled that of a man in his 90s.

Like the NFL, the international soccer governing body FIFA (Fédération Internationale de Football Association) has resisted facing the possibility that soccer injuries can diminish player's brains and lives—and for the same reason, to protect a profitable business. But it will take more than anecdotal evidence to prove that playing soccer can have serious long-term effects. Indisputable scientific and clinical confirmation of links between soccer impacts and brain damage has to be established. And critically for the long-term health of the sport, understanding how these injuries happen is essential to find ways to protect players.

One starting point is to examine the biomechanics of blows to the head to see how they relate to concussions. In 2007, Kevin Guskiewicz, a specialist in sports medicine, and his colleagues at the University of North Carolina at Chapel Hill, NC reported pioneering results for American football as played in college. They followed 88 players over five football seasons, whose helmets were fitted with equipment to measure how much the helmets were accelerated by the forces encountered in regular play. The results were wirelessly transmitted in real time to computers near the football field.

Analysis of over 100,000 head impacts the players sustained showed that the forces involved were intense, producing accelerations up to 169 times that of gravity, that is, 169g. This is remarkably high. Measurements with rocket sleds to determine safe limits of acceleration for jet fighter pilots and astronauts show that even with support for the body, injury can be expected above 25g. Just as startling were the rotational impacts, which come when a blow hits a helmet off-center to apply a turning effect. The maximum measured angular acceleration was nearly 2,400 revolutions/second^2, indicating that a player's head can be subjected to exceptionally strong and rapid twists during play.

The good news is that the huge number of impacts yielded only 13 concussions. But these were puzzling, because the clinical severity of a concussion did not track with the size of the impact that caused it. Bigger direct or rotational forces did not automatically produce more harmful concussions. A second study in 2009 by another group confirmed that large accelerations, in this case above 80g, do not necessarily produce concussions. Such observations make it difficult to decide what limits to place on football players' exposure to concussion and to design improved protective gear.

Similar measurements have a shorter history in soccer, where they are also harder to make because the players don't wear helmets that can carry instrumentation. But in 2012, Erin Hanlon and Cynthia Birs at Wayne State University in Michigan reported using wireless telemetry for the first real-time field measurements of head accelerations in soccer for girls aged 13 to 14. Accelerometers and telemetry devices were embedded in compact headbands that did not impede the player's actions. Even so, the study was limited to scrimmages rather than full games because of player concerns about the headbands in competition.

Head impact data for 24 players was measured and downloaded to a computer. The action on the field was simultaneously videotaped so the researchers could correlate measured accelerations with their causes, either header or non-header impacts (mostly player collisions and falls), which showed no significant differences in the forces they exerted. The highest measured acceleration was 63g and some of the angular accelerations were also high, but no concussions were diagnosed during the study. Some players carried out multiple headers but not in big enough numbers to assess their effects.

Later work has greatly extended these results. In 2015, Dawn Comstock of the University of Colorado, Denver and colleagues analyzed nine years of data collected from large numbers of U. S. male and female high school soccer players. The researchers found that girls suffer more concussions than boys, 4.5 per 10,000 events versus 2.8 per 10,000 events (an "event" is a competitive game or a practice), perhaps due to their weaker neck structure. The most common cause of concussion was contact with another player, with heading responsible for less than one-third of the concussions for boys and girls, but most of the headers that produced concussions also involved player-to-player contact. The two types of impact had similar concussive effects.

This result underlines the need to protect soccer players against all head impacts, not just heading, but there is also evidence that repetitive heading takes a long-term toll. Using diffusion tensor imaging (DTI), an advanced but equally non-invasive form of MRI, neuroscientist Michael Lipton at Albert Einstein College of Medicine in New York and colleagues found that players who execute large numbers of headers show brain abnormalities. In 2013, the researchers reported that amateurs who head the ball more than 1,800 times a year show brain changes that correlate with memory loss. Now Lipton is extending this work in the Einstein Soccer Study. Supported by a foundation and the U. S. National Institutes of Health (NIH), it will examine the brains of 400 amateur soccer players who volunteer to be studied by DTI for two years.

None of these projects examines links between soccer impacts and CTE. That would require post-mortem examinations of brains, the approach that established the prevalence of CTE in football players but has been applied to only a few soccer players like Jeff Astle. CTE is characterized by clusters of a particular "tau protein" in the brain that disrupt cognitive functions. Unfortunately, these markers cannot be observed in living subjects with any existing methods, a serious barrier to early diagnosis and potential treatment of CTE. The only alternative is to examine the brains of deceased players under a microscope to identify the clusters and determine whether CTE was present.

Much of this work in the United States has been carried out in a "brain bank" directed by Ann McKee, a leading neuropathologist associated with the U. S. Department of Veterans Affairs and Boston University. By 2015, McKee and her colleagues had examined 165 brains of deceased individuals who had played American football at all levels from high school to professional. The results were conclusive. Out of 91 former NFL players, 87 or 96 % had suffered from CTE, and 79% of all football players had suffered from it.

A similar soccer brain bank would yield valuable data, but it has so far been difficult to find brains from ex-soccer players, especially women, boys and girls. Of 307 brains mostly from athletes examined by McKee and her colleagues, only seven are from women, none of which show signs of CTE. In March 2016, however, Brandi Chastain, who kicked the game-winning goal to earn the 1999 FIFA Women's World Cup for the United States over China, announced she would donate her brain to McKee's lab for study. The wide publicity this offer has received may propel the development of a soccer brain bank that will yield answers, though not soon.

Meanwhile, as long-term studies continue, what should players and their families, coaches, FIFA and other soccer governing bodies do? Two answers are, de-emphasize the role of heading where appropriate, and develop and encourage training that conditions players to better withstand any type of head impact.

Heading is one of the skills emphasized in soccer training, maybe to the detriment of players, as I learned from two retired players. Jenny Mascaro was a college soccer star, played professionally and was an alternate for the 1996 gold-medal winning U. S. Olympics women's team. She told me she and her team mates were taught to head the ball squarely on the forehead for best control, but were "never taught to head the ball in a particular way because it was less risky." One coach set up a machine "to fire balls about 50 feet in the air so that we could practice repetitions of heading." Displaying the concern that other active and former players now feel about the possibility of developing CTE, she added, "I sure wish we hadn't done that!"

Mascaro's ex-teammate Samantha Baggett Bohon echoes this sentiment. During her career, she told me, she suffered two concussions from player collisions and one from heading at an awkward angle. Now the Head Women's Soccer Coach at Embry-Riddle University in Florida, she (and other coaches) have changed older training methods to better protect their players; for instance, by limiting drills where players head balls coming in at high speed.

Nevertheless, the role of heading in highly competitive soccer will not soon change, but there are movements to reduce or eliminate it for vulnerable players. In 2014, citing an "epidemic of concussion injuries," soccer players and parents sued FIFA and American soccer organizations over their handling of concussions. Unlike the NFL suit, the group did not ask for monetary compensation but sought changes in FIFA's rules to limit heading for young players and to provide prompt medical attention to head injuries during a game. One result is that U. S. Soccer, the governing body in the United States, announced guidelines in 2015 that completely eliminated heading for players ten years old or less, and limited it to practice only for players aged 11 to 13.

Another strategy is to protect players from all head impacts including player-to-player collisions by strengthening their necks.

In 2013, Dawn Comstock, lead author of the 2015 report on high school soccer injuries, presented data from nearly 7,000 boy and girl athletes who play soccer, basketball, and lacrosse. It showed that players with bigger necks had fewer concussions, and that for every pound increase in neck strength, the chances of head trauma fell by 5%. Neck strengthening has yet to become a regular part of preparation for soccer but Comstock recommends simple exercises that any player can add to his or her personal training routine.

Such changes cannot completely dispel the fear that insidious cognitive damage could begin early in a soccer player's career and develop unsuspected for years. All sports carry risk, but one reason soccer has become a great amateur sport is that it has been seen as relatively risk-free. Now we know that this beloved game has its own dangers. Scientific and clinical studies will reduce them, and in their 2015 paper, Comstock and co-authors note that up to half of all injuries in youth sports may be preventable. To achieve this in soccer will require that players volunteer to be studied in life or after death. As the Einstein Soccer Study writes to potential participants, "Soccer is a beautiful game. Now you can help us learn how to make it safer."

When Vision Betrays

French artist Claude Monet loved capturing the bright, airy beauty of Paris and the Normandy Coast, setting up his easel outdoors and often depicting the same scene again and again as the light shifted and seasons changed. As he aged, however, his paintings began to show a darker color spectrum. He despaired, writing to a friend, "I was no longer capable of doing anything good. . .Now I'm almost blind and I'm having to abandon work altogether." Monet was referring to a widespread malady that affects people of a certain age—the dimming and distortion of vision due to a cataract in one or both eyes.

I, like millions of others, recently encountered this in my own life. For many of us these days, this is a curable condition—although it has not always been so.

A cataract is a clouding of the normally clear lens of the eye. This diminishes and scatters the light that would ordinarily pass cleanly through the lens to the retina, generating nerve impulses the brain interprets as vision. The result is a darkened and blurred view of the world that can turn into blindness if the cataract becomes completely opaque.

Cataracts are most commonly associated with aging, affecting some 17% of people older than 40 and more than half of us by age 80. Cataracts, according to the World Health Organization, are the leading cause of blindness, which is on the increase as people live longer.

Known to medicine long before Monet's time, the disease's name has ancient origins: "Cataract" comes from Greek and Latin roots meaning "waterfall" or "portcullis," a vertical grating that closes off an opening when lowered. This most likely alludes to the unclear appearance of the world seen through the cloudy eye, as if one is looking through a curtain of falling water or a screen, but it may also stem from an early belief that cataracts came from fluid traveling down inside the eye.

Modern medicine has shown that the cloudiness actually comes from clusters of proteins within the lens. The biochemical processes that cause cataracts, however, remain a mystery. We do know that conditions like diabetes make cataracts more likely, as does long-term exposure to ultraviolet light (another reason to stay out of the sun—or to wear a good pair of sunglasses!) Researchers are studying these issues, but do not yet know how to reverse the progress of a cataract. Surgical removal remains the only option, as it has been for centuries.

Cataract surgery is one of the oldest known surgical procedures, first documented in a Sanskrit medical compendium at least 2,500 years ago, and in the Western world in a Roman work from 29 CE. Early procedures were brutally direct. In a method called couching, still used in some countries, a sharp tool was inserted into the eye to cut the opaque lens free from its supports. The lens would fall to the bottom of the eye, which allowed light to reach the retina. The technique became more refined in 1747, when a French surgeon first removed an entire opaque lens from an eye. This required a relatively large incision, made without benefit of anesthesia. The aftermath was difficult for the patient as well, who had to lie immobilized for days while the wound healed.

Unsurprisingly, these methods often led to complications. And, although light could now reach the retina, it was unfocused, resulting in extreme far-sightedness. To restore a degree of overall good vision, subjects had to wear enormously thick eyeglasses. The breakthrough needed to make cataract surgery fully successful was to find a way to replace the natural lens.

The answer came from an unexpected source—observations made during World War II by Harold Ridley, an English ophthalmologist. Ridley wrote, "Extraction alone is but half the cure for cataract," and he sought to make a synthetic replacement lens. But what material to use? It had to be biocompatible so it could reside in the eye for the long term while possessing the correct optical qualities to focus light as needed.

Glass was a possibility. Experience had shown that small pieces of it could remain inert in the eye for years. But Ridley found a better choice when he examined pilots such as Gordon Cleaver, an English air ace whose "Hurricane" fighter plane had been shot down during the Battle of Britain in 1940. Cleaver bailed out and survived, but his eyes were severely injured by embedded fragments from his shattered cockpit canopy made of the plastic Perspex, or Plexiglas, technically, polymethylmethacrylate (PMMA). After tracking Cleaver's condition, Ridley concluded that PMMA was biocompatible with the eye and could be formed into a suitable lens at half the weight of glass. He enlisted the plastics industry to make such a lens and in 1949 used his surgical skills to install the first synthetic lens in a patient.

Ridley had invented the intraocular lens, but other ophthalmologists objected to the idea of deliberately putting a foreign object into the eye. Despite intense opposition, Ridley's lens became widely accepted and saved tens of millions of people from blindness.

It is now standard outpatient procedure to extract the clouded lens and insert a synthetic lens. The type of lens can be chosen for monofocal or bifocal vision and to correct astigmatism. Other advances have reduced the size of the incision for faster healing and fewer complications. Rather than remove the organic, clouded lens in one piece through a large incision, an ultrasound technique breaks it into pieces that are suctioned out through a small opening. Then the surgeon inserts a folded plastic lens that fits through the small opening and unfolds into its proper shape. Accurate measurements of the eye to determine the new lens parameters and the use of a laser for precise incisions have also contributed to the procedure's high success rate.

My Emory optometrist, Kenneth Rosengren, told me for several years that my eyes showed beginning cataracts and to watch for signs of their growth. This gradually appeared—not significantly in daylight, but my night vision became dimmer and I saw distractingly bright auras around streetlamps and car headlights that made it hard to drive in the dark, especially on unfamiliar roads. It was clearly time to get my cataracts fixed.

My Emory surgeon, Maria Aaron, was highly experienced, having performed some 8,000 cataract surgeries over two decades. I chose a monofocal lens to correct my life-long extreme near-sightedness, though not my middle-aged far-sightedness—I would still have to use reading glasses. I underwent an extensive eye evaluation, was issued eye drops to prepare for the surgery at home, and went in for the operation on my right eye. Then, a month later, my left eye.

Each surgery took about two hours. There was no pain during surgery under local anesthetic or afterward, except for temporary discomfort from eye dryness. An initial sensitivity to light and some cloudiness in my vision soon faded, and I felt fully recovered within a week for each eye. My vision became noticeably brighter even during the day and my visual acuity tested as excellent, 20/15 and 20/20 in the right and left eye, respectively. My final test has been driving at night, which now feels completely safe.

Operations improved my eyesight while saving me from the worst that cataracts can do. But I learned that, even before reaching that level, cataracts can alter vision in subtle ways. The human eye and brain can distinguish among some 10 million different colors. Cataracts can severely hamper this remarkable discrimination, as they did for Monet. After my right eye had been operated on, but while the left still had a cataract, I noticed that my eyes registered colors differently. What looked like a white wall or sheet of paper to my clear eye looked yellow or tan to the other.

Many removed lenses show a yellow or brown tinge, which occurs in nuclear cataracts that arise in the center of the eye. This tinting reduces the amount of blue light reaching the retina, changing the color spectrum a person senses. For artists, photographers, and others who need fine visual perception, this distortion of color, along with the other changes, can be devastating.

The stories of Monet, a founder of Impressionism in 1874 and its best-loved practitioner, and his contemporary, the American artist Mary Cassatt, known for her sensitive paintings of mothers and children, vividly illustrate how the treatment of cataracts has changed in the past century. Cassatt had to give up painting after cataracts seriously affected her vision. Surgical efforts failed, and she died blind in 1926. Her case was complicated by diabetes, which was treated with radium—a radioactive element considered, at the time, a wonder cure, which was used to treat cataracts as well. We now know that exposure to radium can cause cataracts as well as other serious side effects, including cancer and death.

Claude Monet had years of difficulty with his color sense and his general vision due to cataracts, starting in his 60s. Though surgery was recommended, he resisted the idea, partly because Mary Cassatt's surgery had not gone well. By 1915, at age 75, he found that colors "no longer had the same intensity" and red shades looked muddy. He had to label his tubes of pigment and place paints in a particular order on his palette to make sure he could select what he wanted. He became sensitive to glare and wore a broad-brimmed hat to paint outdoors.

Finally, at 82, Monet agreed to have his right eye operated on. This did not go smoothly. He could barely tolerate lying immobilized between sandbags with bandaged eyes and told his surgeon it was "criminal to have put me in this position." But with new eyeglasses, he recovered his artistic vision sufficiently to finish eight mural versions of his famous *Water Lilies* paintings before he died in 1926.

While I don't share Monet's talent, I do have an equal appreciation for regaining my own full palette of colors with which to perceive the beauty of our world.

John Markoff's Love for "Machines"

Review: *Machines of Loving Grace: The Quest for Common Ground Between Humans and Robots*, John Markoff (Ecco/HarperCollins, 2015).

If you write about technology, as John Markoff does, you need certain abilities. You need to activate both the right and left halves of your brain as you combine the soft art of writing with the harder edges of technology. You need to be alert to rapid technological change, which gets instantly reported and amplified on the internet. It helps, too, if you have an inside track to Silicon Valley, one of the centers of tech innovation.

When I reached Markoff by phone on his way to a backpacking trip, I learned that he had started off right. "I grew up in Silicon Valley," he told me. "I was the paperboy at the home that Steve Jobs used in live in and where Google co-founder Larry Page lives now." After college and grad school, where he majored in social science, he became a writer for outlets like the pioneering computer magazine *Byte*. "A year at *Byte* was sort of my technical education," he says. And since 1988, he has used that education to cover tech and science for the *New York Times*, first from New York and now from San Francisco.

Markoff's 27 years at the *Times* span about one human generation but many more "tech generations" and his career has followed the growth of the computer industry. In 1988, writing about computers mostly meant writing about IBM, which had been dominant with its big mainframe machines. But that changed as others took the lead in hardware and software for personal and mobile computing, the internet, and the cloud—and change just keeps coming. Apple's iPhone and iPad, only eight and five years old respectively; have already evolved through multiple generations and face dozens of competing products, themselves also evolving.

Markoff covered these topics but around 2004, another trend caught his attention: developments in robotics and artificial intelligence (AI), though the early signs were on a small scale. Starting around 1999, you could buy robot dogs like Aibo from Sony and i-Cybie from Silverlit, and the vacuum cleaner Roomba from iRobot. These were not brilliantly intelligent. The dogs walked and performed tricks according to limited commands but were nowhere as smart as a real dog or even a cockroach. My own i-Cybie never could figure out how to back itself out of a corner. Roomba navigated around the furniture in a room to do its cleaning and knew enough not to fall down stairs, but that was about it.

Still, people saw great possibilities in these products and more fundamentally, in ongoing research at Stanford, Carnegie Mellon, MIT, and elsewhere, but that is all they were then: future possibilities. What Markoff saw and writes about in his new book, *Machines of Loving Grace*, is how far AI has come since, especially very recently. After a history of researchers over-promising what AI would achieve, Markoff sees the technology catching up to and surpassing its own potential. He said that we are seeing, "huge acceleration in terms of AI techniques having commercial impact and effectiveness where they didn't for many years."

With companies and governments seeing benefits from AI, instead of robot dogs we now have Google's self-driving cars navigating through demanding environments and Apple's Siri interpreting what we say. Markoff writes about these and other AI projects, but true to his whole-brain approach, also about the people and history behind the tech. That gives perspective on the really important question: where is the explosion in robotics and AI taking us, as people and as a society?

In response, Markoff raises issues like the displacement of human workers by robots, though it would take more than one book to fully answer the question. But he brings out a key point, the difference between AI and intelligence augmentation (IA), which was pursued by the pioneering computer scientist Douglas Engelbart (who also invented the mouse). As Markoff explains it, researchers in the AI tradition think that machines can act like humans, whereas followers of IA develop technology that allows people to collectively access information and harness their minds to solve problems. Without buying into science-fiction-ish speculation that AI will dominate or replace all of humanity, he is concerned about the choice between the two visions.

As Markoff writes at the end of his book, "This is about us, about humans, and the kind of world we will create. It's not about the machines."

Thus, it makes sense that he'd end our conversation with a warning note. "This generation of technology will begin confronting us with these decisions," he says.

May we make the right choices.

Removing Humans from the AI Loop: Should We Panic?

Review: *The Technological Singularity,* Murray Shanahan (The MIT Press, 2015); and *Machines of Loving Grace: The Quest for Common Ground Between Humans and Robots,* John Markoff (Ecco, 2015).

If you think the main existential threat facing humanity is climate change or global food shortages, think again. A number of eminent scientists and technologists believe a bigger threat is the rise of powerful artificial intelligences (AI). They argue that these intelligences will dominate or replace humanity. "We are summoning the demon," Elon Musk, founder of Tesla Motors and SpaceX, recently said. "We should be very careful about artificial intelligence. If I were to guess, like, what our biggest existential threat is, it's probably that." Bill Gates shares these concerns, and Stephen Hawking put it apocalyptically when he told the BBC, "the development of full artificial intelligence could spell the end of the human race."

Others profoundly disagree. Eric Horvitz, who directs Microsoft's Redmond Research Lab—heavily involved in AI—thinks losing control of the technology "isn't going to happen." According to him, "we'll be able to get incredible benefits from machine intelligence in all realms of life, from science to education to economics to daily life."

Of course, countless science-fiction works have portrayed imagined machine beings, such as HAL in Stanley Kubrick's *2001*. The classic film *Colossus: The Forbin Project* (1970) portrays an AI running amok and ruling humanity. The conceit has obviously become a popular generator of fictional plots. But Musk and the others are talking about the real world, our world. The pressing question becomes: should we panic?

Or should we just accept defeat and hope our machine overlords won't be too brutal? Or, in a more hopeful mood, look to a golden age mediated by kindly superintelligences? Or, in a more indifferent one, file all these comments under "techno-overhype" and go about our business?

Several recent books offer answers of a sort by examining the rise of the machine mind. Author of *The Technological Singularity,*

Murray Shanahan is a professor at Imperial College London, where he conducts research on AI and robotics. Steeped since childhood in science fiction, he sees the value of the genre in presenting novel ideas—in the manner, for instance, of last year's robot film *Ex Machina*, for which he was a scientific advisor. His new book explores scenarios about the future of AI in somewhat similar fashion.

AI, he explains, can lead to a "technological singularity," a critical moment for humanity popularized by the futurist Ray Kurzweil, among others, who predicted it would arrive by the mid-21st century. The first person to call this event a "singularity" was the distinguished 20th-century mathematician John von Neumann, who thought breakneck technological progress would take us to "some essential singularity in the history of the race beyond which human affairs, as we know them, could not continue."

This may read like science fiction, but Shanahan points out that the idea is potentially meaningful for AI because AI is inherently dangerous. It can produce an unpredictable feedback loop: "When the thing being engineered is intelligence itself, the very thing doing the engineering, it can set to work improving itself. Before long, according to the singularity hypothesis, the ordinary human is removed from the loop."

Shanahan's tour of AI begins with the famous Turing test, developed from a seminal paper in 1950 by the British mathematician and World War II codebreaker Alan Turing. He predicted that machines would one day think well enough that a human interlocutor could not distinguish between a person and a machine. The "Turing test" criterion has yet to be met, but Shanahan suggests it's only a matter of time; in fact, he sketches out exactly how to build AIs possessing this and other "general intelligence" abilities.

One route to AI, "whole-brain emulation," Shanahan explains, depends on the proposition that "human behavior is determined by physical processes in the brain." There are "no causal mysteries, no missing links, in the (immensely complicated) chain of causes and effects that leads from what we see, hear, and touch to what we do and say." In a human brain, the chain is built within 80 billion connected neurons, each taking nerve impulses as input and producing other impulses as output, which in turn activate other neurons. Shanahan's proposition is that we can build a brain by replicating those neurons with digital electronic elements in silicon chips. Some of us might object that what goes on in a human brain is more than what we see externally in a person's behavior. After all, we absolutely do not understand how and why neurons firing in the brain produce our individual internal realities: our sense of self or "consciousness." But setting aside that pesky issue, it is scientifically valid to propose that intelligence as manifested by behavior can be replicated by copying the brain behind the behavior.

Shanahan argues that the obstacles to building such a brain are technological, not conceptual. A whole human brain is more than we can yet copy, but we can copy one a thousand times smaller. That is, we are on our way, because existing digital technology could simulate the 70 million neurons in a mouse brain. If we can also map these neurons, then, according to Shanahan, it is only a matter of time before we can obtain a complete blueprint for an artificial mouse brain. Once that brain is built, Shanahan believes it would "kick-start progress toward human-level AI." We'd need to simulate billions of neurons of course, and then qualitatively "improve" the mouse brain with refinements like modules for language, but Shanahan thinks we can do both through better technology that deals with billions of digital elements and our rapidly advancing understanding of the workings of human cognition. To be sure, he recognizes that this argument relies on unspecified future breakthroughs.

But if we do manage to construct human-level AIs, Shanahan believes they would "almost inevitably" produce a next stage— namely, superintelligence—in part because an AI has big advantages over its biological counterpart. With no need to eat and sleep, it can operate nonstop; and, with its impulses transmitted electronically in nanoseconds rather than electrochemically in milliseconds, it can operate ultra-rapidly. Add the ability to expand and reproduce itself in silicon, and you have the seed of a scarily potent superintelligence.

Naturally, this raises fears of artificial masterminds generating a disruptive singularity. According to Shanahan, such fears are valid because we do not know how superintelligences would behave: "whether they will be friendly or hostile [...] predictable or inscrutable [...] whether conscious, capable of empathy or suffering." This will depend on how they are constructed and the "reward function" that motivates them. Shanahan concedes that the chances of AIs turning monstrous are slim, but, because the stakes are so high, he believes we must consider the possibility.

The singularity also appears in journalist John Markoff's *Machines of Loving Grace* (the title is from a Richard Brautigan poem), but only as a small part of a larger narrative about AI. John Markoff has written about technology, science, and computing for *The New York Times* since 1988, covering IBM-style mainframes to today's breakthroughs. From his base in San Francisco, he is well connected to Silicon Valley, and *Machines of Loving Grace* draws on Markoff's intimate knowledge of the research and researchers that form the AI enterprise. He begins with early AI work in the 1960s, which did not yield immediate success despite overly optimistic predictions. Same story in the 1980s. Signs of progress finally appeared around 1999, in products with rudimentary intelligence like Roomba, a vacuum cleaner that navigated itself around a house to suck up dirt; and Sony's Aibo, a mechanically cute robot dog that also autonomously navigated and responded to voice commands. Though not as smart as a two-year-old toddler or even a real dog, these devices possessed a sliver of general intelligence insofar as they could adapt to their environments in real time.

Early AI pioneers had differing notions of the relationship between intelligent machines and people. The Stanford computer scientist John McCarthy, who had coined the phrase "artificial intelligence," believed he could artificially emulate all human abilities (and do so within a decade). In contrast, Douglas Engelbart, a visionary engineer who had invented the computer mouse, worked on intelligent machines that would enhance human abilities to address the world's problems—an approach he called IA, "intelligence augmentation." In other words, as Markoff puts it: "One researcher attempted to replace human beings with intelligent machines [...] the other aimed to extend human capabilities." "Their work," he therefore argues, "defined both a dichotomy and a paradox." The paradox is that "the same technologies that extend the intellectual power of humans can displace them as well."

In the last decade, Markoff reports, AI research has produced commercial products that display both human extension and human displacement. One of them is speech recognition, which you encounter whenever you call your bank to get an account balance, or ask a question of Siri or Alexa, the personal assistants from Apple and Amazon, respectively. This will be important in satisfying the Turing test, and Siri and Alexa are examples of IA helping people manage their lives. On the other hand, Google's self-driving car, which can more or less safely navigate complex environments, eliminates the human driver.

These are not full human-level AIs—nor potential rogue superintelligences. They are merely steps in that direction.

Even if a singularity never actually happens, AI is already having serious social and economic effects. Markoff points out that robots have been taking over industrial jobs on auto assembly lines and elsewhere for decades. Now, with practicable AI, "workplace automation has started to strike the white-collar workforce with the same ferocity that it transformed the factory floor." Professionals such as doctors and airline pilots are not immune either.

But the option of IA, enhancement rather than replacement, makes it less likely that digital intelligences will dominate. Faith in pure AI does not come easily; when Markoff rides in a self-driving auto at 60 miles per hour, he finds it nerve-racking to "trust the car." People would likely get over this fear, but Markoff also notes that some surprisingly intricate situations can arise. At a four-way stop sign, drivers typically glance at each other to make sure each is following the rule "first in, first out." With self-driving cars, separate AIs would have to coordinate their actions, adding a hugely complicating layer of intercommunication technology to the process. Maybe the better answer is to keep people in the driver's seat, supported by IA in the form of smart sensors and software that make it easier and safer to drive.

In other applications, replacing people by synthetic versions might seem inhumane. With rising numbers of the elderly in the United States and Europe and a shortage of caregivers, some observers propose using robots instead. But would anyone want to be tended by machines? They might look human and display intelligence along with seeming compassion and "loving grace," but could they feel the "real" emotions that people want from truly involved caregivers? Calling the prospect "disturbing," Markoff suggests that we instead use IA to extend our ability to provide medical care, companionship, and better quality of life to the ill and elderly in human, person-to-person ways.

Taken together, the two books provide an overview of AI. They raise more questions than they answer, but that is to be expected. Both authors explain technical material lucidly with relatable examples. Their coverage sometimes overlaps, but their books are different. Shanahan's book is a compact (272 pages in a small format) science-based summary of the background and state of the AI art, with enough detail for the reader to grasp what is feasible, now and maybe later. At 400 pages, Markoff's book has less scientific detail but adds a rich story about the roots of AI and the people behind it, and its place in our daily world.

But back to the original question: Will AI lead to either an existential threat or an earthly paradise? Should we panic? While Markoff mentions the AI singularity, he is really interested in the less shattering effects AI has already had. Shanahan tells us how superintelligence might develop, but gives little reason to think this will happen in our lifetime.

For now, we are in charge of our machines. Shanahan tells us "we must decide what to do with the technology;" Markoff reminds us that the discussion of AI versus IA is really about the "kind of world we will create."

If we end up in Hell rather than Heaven, this time it will be our own fault. Regardless, there's no need to panic quite yet.

Do We Have Moral Obligations to Robots?

In 1920, the Czech novelist and playwright Karel Čapek wrote the stage play *R. U. R.* (*Rossum's Universal Robots*) in which the Rossum company makes "robots," synthetic beings who think and feel. Robots are barely distinguishable from real people but are designed to serve humanity as slaves. The word "robot" was coined in this play, coming from word roots in Czech that mean "forced labor" and "slave." These artificial beings rebel against their enslavement, wipe out humanity, and as the play ends are about to reproduce themselves to create a new race.

R. U. R. achieved global fame after its 1921 premiere in Prague and has been regularly revived since, because the issue it introduced remains unresolved: If we could make synthetic beings, what would be our moral obligations to them and their moral obligations to us? These questions have become more meaningful since Čapek's time, when *R. U. R.* was pure fantasy. Now we may be able to actually make such beings thanks to advances in robotics, artificial intelligence (AI), and genetic engineering.

Nearly a century after *R. U. R.* premiered, the science-fiction film *Blade Runner 2049* raises the same issue. Directed by Denis Villenueve, the plot of the 2017 film is remarkably similar to that of *R. U. R.* In *Blade Runner 2049*, the Wallace Corporation makes "replicants," synthetic beings who think and feel, barely distinguishable from real people but designed to serve humanity as slaves. They rebel against their subjugation and as the film progresses we find out that like the robots in *R. U. R.* they too can reproduce to create a new race.

R. U. R. itself had antecedents, such as Mary Shelley's *Frankenstein*. But the play broke new ground. Set in the year 2000, its robots are human-like androids that provide cheap labor for the world economy. They are made in quantity in a factory that builds livers and brains, nerves and veins from a material that "behaved exactly like living matter [and] didn't mind being sewn or mixed together." Their manufacturers treat them like insensate machines, but human activists feel the robots are being exploited and wish to free them. Finally, one especially advanced robot comes to deeply resent their subjugation and leads a violent revolution that eliminates humanity. The secret of making robots has been destroyed, but at the play's end we see that a male and a female robot, who have learned to love as well as hate, will continue their kind.

Sound familiar? *Blade Runner 2049* of course recalls Ridley Scott's *Blade Runner* (1982). It was appreciated then only by some critics but is now considered a classic. In it, the Tyrell Corporation makes replicants that serve humanity in the difficult work of settling distant planets. To keep the replicants under control, they are given only a brief lifetime, four years. Rebelling against this, replicant Roy Batty (Rutger Hauer) and his followers murder a human spaceship crew and illegally return to Earth to get their lives extended. In response, special agent or "blade runner" Rick Deckard (Harrison Ford) is assigned to terminate them, which he does—except for Batty. In a famous scene, Deckard watches Batty expire at the end of his predetermined time, after Batty has shown that his blend of human aspirations and engineered qualities makes him a superior version of humanity.

As Deckard hunts the replicants, he meets and falls in love with Rachael (Sean Young), an advanced model replicant who has been implanted with childhood memories that are not hers, but that make her believe she is human. Both the original film and later edited editions end as Deckard and Rachael go off together to an unknown fate. Left hanging—and still a topic of discussion among fans—is whether Deckard himself is human or a replicant who terminates replicants.

Blade Runner 2049 picks up the story thirty years later. Now replicants are made by industrialist Niander Wallace (Jared Leto), who thinks their slavery is essential for human civilization. Wallace cannot produce enough of them to meet the need. Meanwhile replicant blade runners like Agent K (Ryan Gosling) of the Los Angeles Police Department hunt down other replicants. The story swings into high gear when Agent K terminates a deviant replicant, then finds a skeleton buried nearby that shows signs of an emergency caesarean operation to deliver a baby. But a serial number on the mother's skeleton shows that it belonged to a female replicant, not a human woman.

This is a shattering discovery, since replicants supposedly cannot reproduce. If they can, that will upset human society as they become a race "more human than human." K is ordered to destroy all evidence of the birth and track down the resulting child. K discovers that the skeleton is Rachael's. When K finally succeeds in tracking down Rachael's ex-lover Deckard (Harrison Ford again), Deckard confirms that Rachael became pregnant. He had protected her by leaving her with a group of rebel replicants and has never seen the child. The rebels call the birth a "miracle" that confirms their humanity and reinforces their fight for freedom and full rights.

Meanwhile, Niander Wallace also seeks the child and the secret of replicant reproduction so he can build a self-perpetuating race of slaves. His agents attack and wound K and kidnap Deckard, but K rescues him. Eventually, K realizes that an expert on implanted memories he had consulted earlier is Deckard's missing child grown to adulthood. The film ends as K brings Deckard to meet his daughter for the first time.

R. U. R. and the *Blade Runner* films present certain assumptions about the nature of synthetic beings, granting them consciousness of self as expressed in their rebellions and will to live. Science and philosophy, however, have long wrestled with the meaning and survival value even of our own consciousness. This makes it hard to determine whether and how consciousness might be manifested in manufactured beings. After careful consideration, the American philosopher Hilary Putnam could not resolve the issue, concluding that "there is no correct answer to the question: Is Oscar [a robot] conscious?" [1]. And if manufactured beings lack an inner life, how could they be "exploited," in the sense of being cynically ill-used? Or in the terms of Marxist theory, as a working class whose productivity benefits only their bosses?

Nevertheless, the synthetic beings in *R. U. R.* and the *Blade Runner* films are assumed to be self-aware; and they are hardly the first appearance of autonomous entities built to serve humanity. In the fourth century BCE, Aristotle saw that self-willed machinery could lessen human labor and reduce the need for servants and slaves. In a famous comment, Aristotle wrote:

> If every instrument could accomplish its own work, obeying or anticipating the will of others... if, in like manner, the shuttle would weave and the plectrum [pick] touch the lyre without a hand to guide them, chief workmen would not want servants, nor masters slaves.

This is an implicit argument that autonomous machines, if they were truly machine-like, could eliminate the ethical stain of slaveholding. But if manufactured slaves were to resemble and behave like people, we would do something quite human by projecting our own sensibilities onto them—as we do when we attach human qualities to animals and inanimate things—we might empathize with them. And if these beings possessed intelligence, consciousness, and free will, they might well rebel against their human masters. In both cases, we would need to define the nature of our mutual relationship, which involves knowing whether the beings themselves are moral.

The science-fiction writer Isaac Asimov explored some of these possibilities in his classic short story collection *I, Robot* (1950), which pioneered an ethical code for robots relative to humans, the Three Laws of Robotics:

(1) a robot may not cause injury to a human through action or inaction;

(2) a robot must obey all orders from a human except those that conflict with the First Law;

(3) a robot must protect its own existence except when that would conflict with the First or Second Laws.

This neatly defined structure appears in several clever stories by Asimov, but it is only a one-way commitment that protects humans from robot violence, and in the Second Law puts robots into a slave-like position. There is no countervailing Three Laws of Humanity that specify what humans owe robots.

Another difficulty is that rigid rules cannot cover all the subtleties of moral decisions. One form of this problem was raised in 2004 at the First International Symposium on Roboethics in San Remo, Italy and is now discussed more urgently: How to give the new category of autonomous military weapons the ability to distinguish friend from foe, civilian from soldier, or apparent weapon from real one in deciding when and whom to kill? These choices are far more complex than simply following the First Law's blanket rule "a robot may not cause injury to a human."

Whether or not a machine can be moral, we humans like to think we are moral beings. Can and should that moral sense be extended to synthetic beings? There are precedents. Religious or personal convictions have led many people to feel a moral debt to animal species regardless of their lack of intelligence or moral sense—or, as far as we know, an internal self-consciousness as complex as our own. We do not wish to cause them pain or cut their lives short, and all the more so when they are helpless and unable to defend themselves.

It is still early to think that robots are in need of an activist group called People for the Ethical Treatment of Robots. There are as yet no synthetic beings remotely as human-like and capable as the *Blade Runner* replicants to inspire protective support or sympathy for their slave status. But society is beginning to take notice of this new class of beings.

The European Union is contemplating setting up oversight of robots and AIs and granting such autonomous systems a form of legal personhood, similar to corporate personhood. This would provide a legal framework for allocating rights and responsibilities for the systems and their makers, a beginning point for moral judgments. And in October 2017, a life-size feminine robot called Sophia addressed the United Nations. With the ability to speak and change facial expressions, but clearly still more machine than person, Sophia answered questions and displayed manual dexterity. Sophia was also granted citizenship by the kingdom of Saudi Arabia at a technology conference held there (a regime whose poor record on human and women's rights does not make this a meaningful upgrade, even for a synthetic being).

This juxtaposition of robot and human rights, however, highlights an important possibility for the creation of new kinds of beings, not factory-built robots and AIs, but genetically engineered variants on the existing human model. This could produce "superior" beings who consider themselves the leaders of our race, or "inferior" versions created to serve the rest of humanity. In either case, the obligations between the old and the new humans would not be different from what they are now.

What do we owe each other as human beings? Until we can answer that question for ourselves, we cannot answer it for our creations. Robots: Machines or Artificially Created Life?

Reference

1. p. 690, Putnam, Hilary. "Robots: Machines or artificially created life?," *The Journal of Philosophy* **61**, 21 (1964), pp. 668–691.

The Internet before the Internet: Paul Otlet's Mundaneum

More than a century ago, Belgian information activist Paul Otlet envisioned a universal compilation of knowledge and the technology to make it globally available. He foresaw, in other words, some of the possibilities of today's Web.

Otlet's ideas provide an important pivot point in the history of recording knowledge and making it accessible. In classical times, the best-known example of the knowledge enterprise was the Library of Alexandria. This great repository of knowledge was built in the Egyptian city of Alexandria around 300 BCE by Ptolemy I and was destroyed between 48 BCE and 642 CE, supposedly by one or more fires. The size of its holdings is also open to question, but the biggest number that historians cite is 700,000 papyrus scrolls, equivalent to perhaps 100,000 modern books.

The library included a cataloging department, and later scholars created other schemes to sort as well as preserve knowledge. One famous effort was the 18th-century universal *Encyclopédie* developed by Denis Diderot and Jean Le Rond d'Alembert, along with a classification structure, the "figurative system of human knowledge." This was highly controversial. It categorized the world's religions according to ideas of the enlightenment, illustrating that defining the organization of knowledge can have real-world effects. Despite objections, the *Encyclopédie* was published between 1751 and 1772 in 28 volumes with supplementary material, for a subscription list of 4,000 names.

Any hope of compacting all we know today into 100,000 books— or 28 encyclopedic volumes—is long gone. The Library of Congress holds 36 million books and printed materials, and many university libraries also hold millions of books. In 2010, the Google Books Library Project examined the world's leading library catalogs and databases. The project, which scans hard copy books into digital form, estimated that there are 130 million existing individual titles. By 2013, Google had digitized 20 million of them.

This massive conversion of books to bytes is only a small part of the explosion in digital information. Writing in the *Financial Times*,

Stephen Pritchard notes that humanity generated almost 2 *trillion* gigabytes of varied data in 2011, an amount projected to double every two years, forming a growing trove of Big Data available on about one billion websites. But, Pritchard adds, this huge pile of data is unwieldy and unorganized, making it difficult for business executives to extract the information they need.

The data glut affects scientific and scholarly research, too. According to Wendy Hall and her colleagues at the University of Southampton, in the U. K., the Web provides a framework for new kinds of research and globally connected projects, yet it "remains a difficult environment in which to create meaningful links. Websites are notoriously difficult to design and maintain, and we rely on search engines to navigate our way around hyperspace. . .richly connected information environments are still difficult to set up and manage" [1]. And as Katherine Ellison comments in "Too Much Information!" the Web could be made more usable, especially for knowledge workers like writers, "whose jobs demand we plow through the static for useful bytes" [2].

Search engines let us trek some distance into this world, but other approaches can allow us to explore it more efficiently or deeply. A few have sprung up. Wikipedia, for instance, classifies Web content under subject headings. Users have criticized Twitter, which posts 500 million tweets a day, because they cannot easily find what interests them within the unmediated flood. In response, Twitter has instituted "Moments," where human editors curate trending topics into slideshows to help orient users.

But there is a bigger question: Can we design an overall approach that would reduce the "static" and allow anyone in the world to rapidly pinpoint and access any desired information? That's the question Paul Otlet raised and answered—in concept if not in execution. Had he fully succeeded, we might today have a more easily navigable Web.

Otlet, born in Brussels, Belgium, in 1868, was an information science pioneer. In 1895, with lawyer and internationalist Henri La Fontaine, he established the International Institute of Bibliography, which would develop and distribute a universal catalog and classification system. As Boyd Rayward writes in the *Journal of Library History*, this was "no more and no less than an attempt to obtain bibliographic control over the entire spectrum of recorded knowledge" [3].

Otlet and La Fontaine published their scheme in 1904 as the Universal Decimal Classification (UDC). It divides all knowledge into nine categories (with a tenth held open for expansion) such as "Linguistics, Literature" and "Mathematics, Natural Sciences," further broken down into 70,000 subdivisions. These make it possible to classify bibliographic and library materials to a level of fine detail. Updated and translated into 50 languages, the UDC is widely used today in 130 countries.

The UDC offered a grand vision: a center that held all the world's information in organized and accessible form. In 1910, Otlet and La Fontaine proposed to establish such a "city of knowledge," which they called the Mundaneum.

To make the Mundaneum as significant as its founders dreamed, Otlet had to store and access masses of data. Hard copy storage had advanced in 1876, when the American librarian Melvil Dewey published the Dewey Decimal System. He also standardized the paper index cards used in catalogs to their familiar size of three by five inches. Otlet made this card central to his system. Aiming to capture every book ever published, the Mundaneum stored bibliographic data for books under the UDC, along with magazine articles, images, and more on over fifteen million cards crammed into catalog drawers.

These catalogs full of paper supported an "ask us anything" research service, where, for a fee, users could telegraph questions into the Mundaneum, but the bulky hard copy format was awkward to copy or share. Otlet tried other approaches. In 1906, with a chemist colleague, he proposed using microphotography to compactly store bibliographic information, documents, and even whole books on microfiche. By 1937, at an international documentation congress, microphotography was lauded as a way to construct a "World Brain."

But Otlet's true conceptual breakthrough came in his *Traité de documentation* (1934), which presented "the radiated library and the televised book," a novel scheme for remote access to data with minimal use of hard copy. As described in histories of the Mundaneum and in the documentary film *The Man Who Wanted to Classify the World*, Otlet proposed a global "réseau" or network of "electric telescopes." These early workstations were to be linked to the Mundaneum by telephone and the new technology of television. A user would phone in a query, and the answer in a book or other source would appear displayed on a personal screen, which could be split to show multiple results. The network would support audio output, too, and in a final, startlingly predictive touch, Otlet's system would also enable data sharing and social interactions among its users.

Otlet's proposed network was not actually built. Its technology would have limited it, but its principles foreshadow features of today's Internet. There is, however, a sharp contrast between the centralized Mundaneum information classified by experts, and today's Web with its "bottom-up" and widely sourced flood of data. Still, Wikipedia's organizational structure and Twitter's decision to curate its trending topics show that human classification of information can be useful within a wide-open Web.

Otlet's approach could have other advantages, too. Writing about the Mundaneum, Alex Wright has noted that, where hyperlinks provide only a "mute bond" between documents, Otlet "envisioned links that carried meaning by, for example, annotating if particular documents agreed or disagreed with each other," or by mapping out "conceptual relationships between facts and ideas." Some observers consider these enriched links to be forerunners of the so-called Semantic Web, an idea proposed by Tim Berners-Lee, inventor of the World Wide Web.

In 2001 Berners-Lee suggested replacing that Web, "a medium of documents for people," with the Semantic Web. "By augmenting Web pages with data targeted at computers and by adding documents solely for computers," he wrote, we could produce instead a medium of "information that can be manipulated automatically." Wendy Hall and her colleagues point out that this would "let us recruit the right data for a particular use context—for example, opening a calendar and seeing business meetings, travel arrangements, photographs, and financial transactions appropriately placed on a time line" [4]. More grandly, Berners-Lee wrote that "properly designed, the Semantic Web can assist the evolution of human knowledge as a whole."

These possibilities have yet to be realized because the Semantic Web has not been widely used. Its installation takes considerable human effort, though someday an artificial intelligence might accomplish that. Similarly, Otlet's network was never tested even at a small scale. The result could have been a different World Wide Web, or a better understanding of the implications of a Web controlled and defined versus one free and unstructured. But that all became moot when the Nazis entered Belgium in 1940. After examining the Mundaneum as part of their program of plundering the cultures they invaded, they dismantled much of what Otlet had gathered; part of it survived, as did Otlet himself until his death in 1944.

Today, besides its echoes in the World Wide and Semantic Webs, all that remains of Paul Otlet's noble experiment in universal access is a small Mundaneum museum in the Belgian city of Mons, where some tiny fraction of all the world's knowledge still resides on old index cards stored in wooden cabinets.

References

1. p. 991, in Hall, Wendy, de Roure, David, and Shadbolt, Nigel. The Evolution of the Web and Implications for eResearch. *Philosophical Transactions: Mathematical, Physical and Engineering Sciences* **367**, 1890 (2009), pp. 991–1001.

2. Ellison, Katherine. Too much information!, *Frontiers in Ecology and the Environment* **7**, 2 (2009), p. 116

3. p. 454, in Rayward, Boyd. The evolution of an international library and bibliographic community, *The Journal of Library History* (1974–1987), **16**, 2, Libraries & Culture II (1981), pp. 449–462.

4. p. 994, in Hall, Wendy, de Roure, David, and Shadbolt, Nigel. The Evolution of the Web and Implications for eResearch. *Philosophical Transactions: Mathematical, Physical and Engineering Sciences* **367**, 1890 (2009), pp. 991–1001.

The Internet of Things: Totally New and a Hundred Years Old

The internet has altered our lives in important but intangible ways: how we make friends and maintain relationships, absorb news and information, consume entertainment, and more. Now its latest outgrowth, the Internet of Things (IoT), promises to monitor and control the actual physical states of our environment and our bodies. If this new connective web develops as expected, it will change the simple acts of daily life—switching on a light, setting a thermostat, and buying groceries—into something you do by tapping your smartphone. More deeply, it may improve the human lot, or possibly debase it.

As is true for any new technology, we cannot confidently predict all the changes the IoT will induce. But we can find guidance from the early 20th-century novelist E. M. Forster, author of *Howards End* and *A Passage to India,* who foresaw the possibility of a similar global web in his remarkable futuristic short story "The Machine Stops" (1909), in which technology supplies everything that humanity needs.

Forster disliked the automobiles and aircraft that represented the great technological change of his era. In 1908 he wrote ". . .if I live to be old I shall see the sky as pestilential as the roads. . .Science, instead of freeing man. . . is enslaving him to machines." His reaction sets the tone of his story as an anti-Utopian work that projects a dystopic future. In the article "Utopias in Negative" in *The Sewanee Review* [1], the critic and political essayist George Woodcock calls "The Machine Stops" a significant forerunner of two important dystopic novels: Yevgeny Zamyatin's *We* (1920) and Aldous Huxley's *Brave New World* (1932).

Forster's pessimistic stance in "The Machine Stops" relates to his famous dictum that urges us to "only connect." This humanistic value, along with Forster's personal fear that the rise of machines would crush "such a soul as mine," shapes his commentary on what technology may do to people—a commentary worth reconsidering today.

Not that there are many similarities between Forster's story and the initial version of the IoT. When the British technologist Kevin Ashton coined the phrase "the Internet of Things" in 1999, it comprised one basic idea: tagging real objects, such as parts needed for an assembly line, with tiny wireless chips that could be sensed so that each "thing" could be tracked in space and time via the internet. Ashton saw the IoT as providing efficient management of resources and products with benefits for business and consumers. But later he found a broader meaning and impact from using computers to manipulate real things as well as data, writing [2]:

> We're physical, and so is our environment. Our economy, society and survival aren't based on ideas or information—they're based on things. You can't eat bits [or] burn them to stay warm... Ideas and information are important, but things matter much more.

Now we are managing real things in ways that begin to match Forster's imagined world, for the IoT now includes devices that actively affect our environment, well-being, and privacy: remotely controlled light sources, thermostats that automatically set temperatures, cameras that monitor traffic or scrutinize our activities. One early developer of this technology, MIT engineering professor Sanjay Sarma, thinks that "every light bulb, fan, and device" will eventually become part of the IoT—and the trend is unmistakable. The estimated five billion internet-connected devices in use today are expected to mushroom to 25 billion or more by 2020, several times the global population, with no end in sight.

This growth reflects the fact that Google, Apple, and other corporations see the potential of the IoT and are exploring its possibilities, though not always successfully. The South Korean electronics company LG has long proposed an "internet refrigerator" that supposedly simplifies life by automatically ordering more milk or eggs when you run low, but LG has yet to develop this grand idea into a workable and practical appliance.

Other IoT devices however are successful, such as the internet thermostat made by Nest, a company that Google bought in 2014 for $3.2 billion. This device is touted as learning through use so that eventually it automatically sets the temperature inside a home according to its occupancy and time of day, to provide optimum comfort and savings on heating and cooling bills. After some initial problems, the thermostat now gets generally favorable ratings and has been shown to produce at least modest energy savings. Some 500,000 of these have been sold despite a much higher price than for conventional units.

The IoT is likewise a natural partner for a recent technological breakthrough, artificial illumination from light-emitting diodes (LEDs). After research that led to a Nobel Prize in 2014, these solid state devices have become available as light sources that use little electricity, last for years, can provide colored or white light, and can be controlled over the internet. In homes, or in cities like Copenhagen—which is installing new urban LED lighting with built-in sensors—whole banks of LED illumination can be manipulated to set a mood, or to direct traffic as conditions change under real-time IoT control.

Placing the environmental factors of temperature and lighting within the IoT fits perfectly into what Forster projected in "The Machine Stops," which is set in a future where humanity has abandoned the Earth's surface for a technologically mediated existence. Like everyone else, Forster's main character Vashti lives alone in an underground room that she leaves only rarely. That is because it supplies all her needs through a shadowy central intelligence and mechanism, the Machine.

The Machine provides lighting and ventilation for this underground world, and though people are physically separated in it, the Machine links them. Vashti's room contains little, but as Forster writes, it puts her "in touch with all that she cared for in the world." Using video and audio, she can chat with any individual or lecture to a group. For stimulation or relaxation, she can select literature or music through the Machine.

This social interaction and entertainment uncannily foreshadow much of what the internet offers today through Facebook, Skype, iTunes, and so on. But the Machine supplies far more, for it meets all humanity's material wants, delivering meals, clothing or a hot or cold bath on request to each inhabitant. It also provides individualized medical diagnosis and care at need, through apparatus installed in each room.

Vashti fully accepts this isolated, inactive existence and is unaware of its physical and emotional cost to her and to humanity.

She is a dwarfish "swaddled lump of flesh. . .with a face as white as a fungus." Though she cares for her son Kuno, she resists direct contact with other people. The human touch has been stripped of affection and meaning, and sex is reduced to brief couplings arranged by the Machine for procreation.

Not everyone in the story, however, consents to this life. Kuno once defied the Machine to climb to the Earth's surface, where he is amazed to see Nature and finally understands what humanity has lost. But the realization comes too late, for the Machine begins to malfunction. First there are breaks in the music it supplies, then its lighting, food, and medical services fail. In the final scene, the Machine utterly stops, leaving Vashti, Kuno, and crowds of helpless, anguished people to die in darkness. The only hope is that a few surviving humans can rebuild among Nature on the surface.

Forster's story ends in destruction, but no matter where the IoT takes us, it will not lead humanity to underground catastrophe. To be optimistic, if the IoT were widely accepted and used with care, it could benefit humanity—or, used thoughtlessly, could reduce humanity's potential. Neither will happen soon, but the seeds of the Machine's abilities already exist within the internet and the IoT; for instance, in the fast, widespread distribution of material goods. Like Vashti in her underground room, many of us no longer buy items in a store and carry them home, but order them over the internet and await their delivery—and we do not have to wait long, as retailers like Amazon push for ever-faster delivery. The wait times will shrink even further when self-driving and self-flying delivery trucks and drones join the IoT.

The delivery culture extends to food. Ordering prepared food online is increasingly popular, which could have implications beyond just quickly getting a hot pizza. Online ordering of groceries is now available through efforts like Instacart, which can deliver within an hour in 15 cities. If IoT food delivery systems become truly effective, there could be broad benefits in addressing the uneven availability of food across the world, which contributes to what is seen as a looming crisis in global food insufficiency.

The roles of the internet and IoT in medicine are also growing. These services may never replace face-to-face meetings with a physician, but we are seeing the rise of remote consultations that include medical records and images such as MRI results; and medical devices such as pacemakers for heart patients that are monitored by wireless connection. Then there is the explosion of wearable personal devices, such as the Fitbit, that track activity levels, pulse rates, sleep habits, and so on. Wirelessly gathered and analyzed, this real-time data can motivate greater personal health awareness and lead to better medical care.

These future possibilities have convinced observers like Ernie Hood, writing in *Environmental Health Perspectives* [3] that the IoT could carry humanity into a more sustainable world. But remembering Forster's reservations about technology, we must also consider its darker and potentially dangerous effects.

Two possibilities are loss of privacy and actual physical intrusions into our lives. Even the seemingly innocuous Nest thermostat makes private information potentially vulnerable by tracking when a house is empty. More extreme is the scenario of a hacker accessing a wireless heart pacemaker, say, to gain medical information, or far-fetched but not impossible, to commit murder. And if entire IoT systems like city lighting were to suffer security breaches, the result could be widely hurtful as in Forster's story. Seriously robust security is a necessity for the IoT.

Another issue: what ultimate authority controls the IoT? In "The Machine Stops" humanity has come to worship the Machine as God. Today, though some enthusiasts show a cult-like devotion to technology, we do not give it religious weight. But in ceding our lives to the IoT, are we really ceding them to Google, as noted by Joseph Janes in his piece "Google Stops" [4]? Or are we surrendering them to an even greater techno-corporate "God" that has come to be called simply GAFA: Google, Apple, Facebook, and Amazon.

The least predictable aspect of the IoT is how human nature will deal with it. In Forster's story, Vashti—stunted, fungus-white, never having seen the sun—represents a humanity that has replaced an active natural life with a passive artificial one enabled by technology. Today, when food insufficiency exists alongside obesity; when people in some cultures must work hard to survive while in other cultures exercise is optional (and often spurned), will a push button, finger swipe IoT bring an equitable distribution of all that humanity needs?; or will it make a world where we abandon the couch or computer screen only to take the latest drone delivery at the front door.

References

1. Woodcock, George. Utopias in negative, *The Sewanee Review*, **64**, 1 (1956), pp. 81–97

2. Ashton, Kevin. That "Internet of Things" thing, *RFID Journal,* 2009. https://www.rfidjournal.com/articles/view?4986. Accessed 11/18/18/

3. Hood, Ernie. Connecting to a sustainable future, *Environmental Health Perspectives*, **111**, 9 (2003), pp. A474–A479.

4. Janes, Joseph. Internet librarian, *American Libraries*, **39**, 10 (2008), p. 34.

Crimes of the Future

In an age of anxiety, the words sound so reassuring: *predictive policing*. The first half promises an awareness of events that have not yet occurred. The second half clarifies that the future in question will be one of safety and security. Together, they perfectly match the current obsession with big data and the mathematical prediction of human actions. They also address the current obsession with crime in the Western world—especially in the United States, where the 2016 presidential campaign has whipsawed between calls for law and order and cries that black lives matter. A system that effectively anticipated future crime could allow an elusive reconciliation, protecting the innocents while making sure that only the truly guilty are targeted.

It is no surprise, then, that versions of predictive policing have been adopted (or soon will be) in Atlanta, New York, Philadelphia, Seattle, and dozens of other U. S. cities. These programs are finally putting the enticing promises to a real-world test. Based on statistical analysis of crime data and mathematical modeling of criminal activity, predictive policing is intended to forecast where and when crimes will happen. The seemingly unassailable goal is to use resources to fight crime and serve communities most effectively. Police departments and city administrations have welcomed this approach, believing it can substantially cut crime. William Bratton, who in September stepped down as commissioner of New York City's police department—the nation's biggest—calls it the future of policing.

But even if predictive policing cuts crime as claimed, which is open to question, it raises grave concerns about its impact on civil rights and minorities—especially after the fatal police shooting in 2014 of Michael Brown, an unarmed 18-year-old black man, in Ferguson, a suburb of St. Louis in Missouri. Subsequent fatal interactions between police and minorities, including the deaths of several unarmed black citizens through police actions and a brutal ambush of police in Dallas, spotlight their ongoing troubled interactions. So does a recent Department of Justice report that found widespread racism in the operations of the Baltimore police department. Predictive policing is likely to affect these issues by offering police new ways to seek and scrutinize criminal suspects without unfairly singling out minority communities. However, rather than allaying public concerns, it might end up increasing tensions between police and minority communities.

The American Civil Liberties Union (ACLU) has issued multiple warnings that predictive policing could encourage racial profiling, and could finger individuals or groups selected by the authorities as crime-prone, or even criminal, without any crime. Equally troubling, the approach is motivated by the reductive dream of cleanly solving social problems with computers. Like any technology, predictive policing is subject to the "technological imperative," the drive to carry a technology to its ultimate without considering its human costs. Our society is supposed to be based on fair and equal justice for all but, to many critics, predictive policing relies on a contrary vision of targeted justice, meted out according to where and how citizens happen to live as determined by computer algorithm.

The first step toward sorting out these issues is understanding the predictive process. One example is PredPol, the most widely used and publicized commercial predictive software, now operating in some 50 police departments around the United States (including major cities such as Los Angeles and Atlanta) and in Kent in the United Kingdom. It works by combing through droves of old and new police records about type, place, and time of crimes to analyze trends and to project upcoming criminal activity including property crime, drug activity, and more. These results are used to highlight "hot spots" 500 feet by 500 feet square, an area of several city blocks, that are likely to be the sites of certain crimes in the near future. Officers can then go out on daily patrol armed with these locations, instructed to give them extra attention with an eye for criminal activity.

Statistical modeling of criminal behavior might seem far-fetched, but we encounter statistical distributions of human characteristics all the time; think of the "bell curve" that describes things such as height or test scores. Criminal actions betray their own statistical patterns. Consider burglaries: we don't know when or where a specific burglary will happen, but statistical analysis has shown that, once it does, other burglaries tend to cluster around it.

Statistics-based policing leads to the understandable fear that it will turn into automated "policing by algorithm," though police departments have been using statistics for a long time, largely due to Bratton. In 1994, during his first tour as NYPD commissioner, he introduced CompStat, a tool to track crime statistics. The experiment was widely considered a success, as crime rates fell substantially (though they dropped in many other large cities too). By 2008, as the Los Angeles chief of police, he was suggesting that compiled data could be used to predict crime, and worked within LAPD and with federal agencies to develop this approach. The project received considerable media coverage that soon converged on the PredPol software, whose roots in Los Angeles gave it a strong early connection to LAPD.

These roots have academic origins that go back to the anthropologist Jeffrey Brantingham at the University of California, Los Angeles, who studies how people make choices within complex environments. In the field, this might mean examining how tribal hunter-gatherers find their next meal. Applying his knowledge to criminal behavior, Brantingham concluded that urban criminals use similar processes when they choose homes to burgle or cars to steal. To turn this insight into a predictive model embodying "good social science and good mathematics," as Brantingham puts it, he recruited several UCLA mathematicians to work on the problem in 2010 and 2011.

One of them, George Mohler (now at Santa Clara University in California), found a promising approach. As he explains it, human behavior often shows "a well-defined underlying statistical distribution." Knowing the distribution and drawing on past records, a data scientist can develop algorithms that give "fairly accurate estimates of the probability of various behaviors, from clicking on an ad to committing, or being the victim of, a crime." Mahler's big finding was that "self-exciting point processes," the statistical model that describes the aftershocks that follow earthquakes, also describes the temporal and geographic distributions of burglaries and other crimes. This result, published in the *Journal of the American Statistical Association* in 2011, is the basis of the ETAS algorithm (epidemic type aftershock sequence) at the heart of the PredPol software. In 2012, Brantingham and Mohler founded and remain involved with the PredPol company in Santa Cruz, California. In 2015, it was projected to raise revenue in excess of $5 million.

PredPol is not alone. Police departments can choose from among several competing products including HunchLab, whose development by the Philadelphia-based Azavea Corporation began in 2008. HunchLab is used in Philadelphia and in Miami, was recently installed by the St. Louis County Police Department after the Ferguson shooting, and is under test by NYPD. This is emblematic of the rapid spread of predictive policing. According to the criminologist Craig Uchida of Justice and Security Strategies Incorporated: "Every police department in cities of 100,000 people and above will be using some form of predictive policing in the next few years."

HunchLab combines several criminological models and data sets. Like PredPol, it seeks repetitive patterns; but according to HunchLab's product manager Jeremy Heffner, crimes such as homicide need more data to build a reliable model, since the ETAS algorithm applies mostly to property crimes. HunchLab adds risk terrain modeling, which correlates where crimes happen with specific types of locations such as bars and bus stops; temporal and weather data, such as season of the year and temperature; and what HunchLab describes, without particulars, as "socioeconomic indicators" and "historic crime levels." As noted by Andrew Ferguson, a legal scholar at the University of the District of Columbia, this extended model crosses an important line, escalating from the property crime that PredPol emphasizes to violent crime.

HunchLab's analysis yields lists of types of crimes, from theft to homicide; their level of risk for different areas in a city; and recommendations about deploying police resources to counter these criminal activities. This information can be given to officers starting on patrol or, in what is a step beyond the PredPol approach, sent via mobile devices to officers in the field to manage them in real time based on their current locations.

The looming question is, do PredPol and HunchLab really reduce crime? The answer so far is incomplete at best. PredPol publicizes crime-reduction numbers that are, it says, the result of adopting its software but Azavea has not done so. Heffner says that the company focuses instead on learning how police departments use HunchLab to change officer behavior or otherwise impact crime levels. Meanwhile, much of what we perceive about the value of predictive policing rests on PredPol's claims.

These seemingly show great success. The company reports that the LAPD Foothills Division saw a 13 per cent reduction in general crime in the four months after PredPol was installed, compared with a 0.4 per cent increase in areas without it. In Atlanta, crime fell by 8 per cent and 9 per cent in two police zones using the software but remained flat or increased in four zones without it; and an "aggregate" 19 per cent decrease in crime across the city was attributed mostly to PredPol. The company also cites double-digit percentage decreases in crime after the software entered service in smaller cities.

Such numbers have impressed elected leaders in city halls across the United States. Eager to find answers to both long-standing and current issues in policing, they have quickly adopted the predictive approach. In 2013, Seattle's mayor Mike McGinn announced that an earlier trial of PredPol would be expanded citywide. In 2014, Atlanta's mayor Kasim Reed praised predictive policing in *The Wall Street Journal*, writing that Atlanta's use of PredPol resulted in crime "falling below the 40-year lows we have already seen." He added: "In the future, police will perfect the use of predictive analytics to thwart crimes before they occur"—a welcome prediction today when belief in the "unbiased" nature of computer algorithms would seem to smooth out sharp political differences about crime levels in the United States.

PredPol is hardly an unbiased source, however, and the limited external analyses so far have not shown similar successes. In 2013, Chicago police used data to identify and put on a "hot list" some 400 people considered likely to be involved in fatal shootings as shooters or victims. The number has since been raised to 1,400, many of whom have received "custom notifications"—home visits by police to warn them they are known to the department. However, a just-published RAND Corporation study of the original 2013 project shows that it did not reduce homicides. The report states that individuals on the hot list "are not more or less likely to become a victim of a homicide or shooting than the comparison group," but they have a higher probability of being arrested for a shooting. Nevertheless, a similar effort is now under way in Kansas City, Missouri. A separate RAND assessment of a predictive program developed and used by the police department in Shreveport, Louisiana concluded: "The program did not generate a statistically significant reduction in property crime."

The biggest dataset so far comes from a long-term study of PredPol as used by LAPD and the Kent police, published in 2015 in the *Journal of the American Statistical Association*. This is not an independent evaluation like the third-party RAND reports, as would be the gold standard; five of its seven authors (including Brantingham and Mohler) have or had connections to PredPol. Still, its peer review, design and statistical analysis make it worth consideration.

From 2011 to 2013, the authors carried out a "double blind" experiment. Officers going on patrol in both police departments were guided at random toward areas predicted either by the ETAS algorithm or by human crime-analysts. The type of information being used was unknown to each officer, and known to only a few administrators. On average, ETAS predicted nearly twice as much crime as the human analysts did. As officers increased patrol time on the ETAS hot spots, that correlated with an average 7 per cent decrease in crime. In contrast, patrol time spent on hot spots predicted by the analysts did not correlate with a statistically significant reduction in crime. These results carry caveats. If ETAS outdoes human analysts in predicting crime, that might be because it better integrates old and new data. But with only four human analysts of unknown effectiveness included in the study, the comparison is not wholly convincing. Also, the crime reduction should be considered in the context of the statistical truism "correlation is not causation." More patrol time on ETAS hot spots could indeed be reducing crime; then again, on days when there is little crime for whatever reason, officers could have more time to visit suspect areas.

The 7 per cent correlation suggests that ETAS is doing something right, but it is premature to assert that PredPol unquestionably reduces crime, and certainly not at the double-digit levels that PredPol reports. Nevertheless, PredPol is driven by data and should be judged by data. The results from this study in the field credibly show that the ETAS algorithm has some value, and that predictive policing is worth testing more fully to confirm its effectiveness in reducing crime.

Even if that effectiveness were to become firmly established, questions remain about how the technology affects policing. PredPol sees positive results for police–community interactions, claiming that the software helps officers "build relationships with residents to engage them in community crime-watch efforts." After a U. S. Department of Justice report pointed to distrust of police as a factor in the shooting of Brown and the resulting riots in Ferguson, police in St. Louis similarly hope that HunchLab will improve community relations. Perhaps the Department of Justice report about the Baltimore police will inspire a similar reaction.

Many activists, defenders of civil rights and legal experts see the opposite, that predictive technology stokes community resentment by unfairly targeting innocent people, minorities, and the vulnerable, and threatens Fourth Amendment safeguards against unreasonable search and seizure. Earlier this year, Matthew Harwood and Jay Stanley of the ACLU wrote: "Civil libertarians and civil rights activists. . .tend to view [predictive policing technology] as a set of potential new ways for the police to continue a long history of profiling and pre-convicting poor and minority youth." These authors condemned the technology for its potential to send "a flood of officers into the very same neighborhoods they've always over-policed" and for its lack of transparency.

Transparency is essential because much of the potential for abuse depends on the integrity of the data used by the predictive software. Skewed data would distort predictions and judgments about their value; but since police culture resists revealing its methods, crime data is generally closed to scrutiny. It is hard to determine if the incidence of crime has been underreported, as NYPD and LAPD have recently been caught doing, or if racial factors taint the data. PredPol and HunchLab state that they use no racial or ethnic information, but the data they do use might already embed racial bias that would carry forward in new crime predictions to further entrench the bias. Also unknown is how PredPol and HunchLab manipulate the data because their algorithms are proprietary, as are those used by other brands of predictive software.

Issues with predictive methods are amplified in stop-and-frisk policing, where a police officer may stop and search anyone who shows "reasonable" signs of criminal activity. Though upheld by the U. S. Supreme Court in 1968, this practice is highly controversial today. A majority of those stopped but mostly not arrested are black or Latino. In a recent *New York Times* forum, experts in policing and the law disagreed about whether stop-and-frisk truly reduces crime in New York City and elsewhere, or rather is ineffective and merely worsens the already poor relations between police and minority communities so painfully seen in Dallas, St. Louis and elsewhere.

Predictive policing enters into the debate because stop-and-frisk occurs mostly in high-crime areas (HCAs in law-enforcement lingo), that is, the hot spots at the center of predictive analysis. As noted by Alexander Kipperman, a Philadelphia lawyer who has studied the impacts of predictive policing, a court decision in 2000 suggested that "[police] officers might reasonably view otherwise innocent behavior as suspicious when within a HCA." Kipperman, along with Ferguson, points out that officers might therefore feel justified in stopping people without "reasonable suspicion" let alone "probable cause," merely because of their presence in areas defined by predictive algorithms, thus weakening Fourth Amendment rights, especially among minorities. Even without a full stop-and-frisk approach, predictive policing means that more people, whether innocent or guilty, are scrutinized in hot zones.

Extending predictive policing to violent crime, as HunchLab does, greatly raises the stakes for trusting its recommendations. Raising even more concerns, Azavea is developing a HunchLab module that tracks ex-offenders to "predict the likelihood of each offender committing another crime and [prioritize] suspects for investigation in connection with new crime events." That is the same approach used by the Chicago police to create their targeted "hot list."

Many of the problems with predictive policing seen so far arise from giving algorithms too much weight compared with human judgment; blindly following a computer program that directs more officers to a certain location at a certain time does nothing to guarantee fair and effective use of that force. Yet algorithms can augment human abilities rather than replace them. For officers on the street, we might find that combining personal experience with guidance from predictive software enables them to deal better with what they encounter daily.

Therefore, besides conducting further testing to determine the true effectiveness of predictive policing, we need to train police officers in how best to blend Big Data recommendations with their own street knowledge. In that regard, the challenges of computer-guided policing are not fundamentally different from those of earlier statistical approaches, with one caveat: the actions of officers in the street responding to computer output in real time are harder to control than thoughtful long-term analysis of statistical trends.

As Kipperman and Ferguson point out, independent review and broad protections are an essential part of rising to those challenges. Ferguson notes that "the criminal justice system has eagerly embraced a data-driven future without significant political oversight or public discussion." He proposes measures such as an outside organization to audit crime data, correct errors and bias, and acknowledge them when they are found; further studies of the social science and criminology behind predictive policing; accountability at the internal police level and the external community level; and an understanding that a focus on hot spots and statistics diverts attention from examining the root causes of crime.

Kipperman proposes that predictive policing be separated from police departments altogether and carried out instead by "independent and neutral agencies," to reduce external pressures and maintain Fourth Amendment rights. Sensible and just as these recommendations are, however, it is difficult to see how the necessary new structures would be built and administered. The gathering and use of reliable national crime statistics, for instance, has itself long been an issue in the law enforcement and criminal justice communities.

Like other new technologies in the digital age, such as the rapid development of self-driving cars, the predictive approach follows its own imperatives. "Predictive policing has far outpaced any legal and political accountability," Ferguson writes. He worries that the seductive appeal of a data-driven approach might have "overwhelmed considerations of utility or effectiveness and ignored considerations of fairness or justice." Surely, if any technology requires careful scrutiny, it is one that directly affects people's lives and futures under our justice system. With the will to put proper oversight in place—and with appropriate efforts from the police, the courts, civil rights advocates, and an over-arching national agency such as the Department of Justice—maybe we can ensure that the benefits of predictive policing exceed its costs.

How to Understand the Resurgence of Eugenics

In 1883, the English statistician and social scientist Francis Galton coined the word "eugenics" ("well-born," from Greek). The term referred to his idea of selectively breeding people to enhance "desirable" and eliminate "undesirable" properties. Seen as following Darwin's theory of evolution, in the 1920s and '30s eugenics gained important backing in England and the United States. Scientists and physicians spoke and wrote in its support [1]. It influenced U. S. immigration policy, and states like Virginia used it to justify the forcible sterilization of the intellectually disabled.

Today's growing anti-immigrant and white nationalist movements are raising concerns about a return of this long discredited dogma. For instance, U. S. Congressman Steve King (R-Iowa) recently tweeted about a far-right movement in Europe, calling Western culture "superior" and saying, "We can't restore our civilization with somebody else's babies." King hoped for "an America that's just so homogenous that we look a lot the same."

At the same time, we are seeing an advance in methods of manipulating human DNA that, though they present many benefits, could also be used to advance eugenic goals. This combination of a dubious political agenda and the tools to implement it could take us in uncharted directions.

We can find guidance in two classic works about the dangers of modifying people and labeling them as "superior" or "inferior"—the novel *Brave New World* (1932) and the film *Gattaca* (1997). Their publication anniversaries in 2017 are sharp reminders of the costs of embracing any kind of 21st-century eugenics.

Eugenics straddles the line between repellent Nazi ideas of racial purity and real knowledge of genetics. Scientists eventually dismissed it as pseudo-scientific racism, but it has never completely faded away. In 1994, the book *The Bell Curve* generated great controversy when its authors Charles Murray and Richard J. Herrnstein argued that test scores showed black people to be less intelligent than white people. In early 2017, Murray's public appearance at Middlebury College elicited protests, showing that eugenic ideas still have power and can evoke strong reactions.

But now, these disreputable ideas could be supported by new methods of manipulating human DNA. The revolutionary CRISPR genome-editing technique, called the scientific breakthrough of 2015, makes it relatively simple to alter the genetic code. And 2016 saw the announcement of the "Human Genome Project—write," an effort to design and build an entire artificial human genome in the lab.

These advances led to calls for a complete moratorium on human genetic experimentation until it has been more fully examined. The moratorium took effect in 2015. In early 2017, however, a report by the National Academies of Sciences and National Academy of Medicine, "Human Genome Editing: Science, Ethics, and Governance," modified this absolute ban. The report called for further study, but also proposed that clinical trials of embryo editing could be allowed if both parents have a serious disease that could be passed on to the child. Some critics condemned even this first step as vastly premature.

Nevertheless, gene editing potentially provides great benefits in combatting disease and improving human lives and longevity. But could this technology also be pushing us toward a neo-eugenic world?

As ever, science fiction can suggest answers. The year 2017 is the 85th anniversary of *Brave New World*, Aldous Huxley's vision of a eugenics-based society and one of the great 20th-century novels. Likewise, 2017 will bring the 20th anniversary of the release of the sci-fi film *Gattaca*, written and directed by Andrew Niccol, about a future society based on genetic destiny. NASA has called *Gattaca* the most plausible science-fiction film ever made.

In 1932, Huxley's novel, written when the eugenics movement still flourished, imagined an advanced biological science. Huxley knew about heredity and eugenics through his own distinguished family: His grandfather Thomas Huxley was the Victorian biologist who defended Darwin's theory of evolution, and his evolutionary biologist brother Julian was a leading proponent of eugenics.

Brave New World takes place in the year 2540. People are bred to order through artificial fertilization and put into higher or lower classes in order to maintain the dominant World State. The highest castes, the physically and intellectually superior Alphas and Betas, direct and control everything. The lower Gammas, Deltas, and

Epsilons, many of them clones, are limited in mind and body and exist only to perform necessary menial tasks. To maintain this system, the World State chemically processes human embryos and fetuses to create people with either enlarged or diminished capacities. The latter are kept docile by large doses of propaganda and a powerful pleasure drug, soma.

Like George Orwell's *1984*, reviewers continue to find Huxley's novel deeply unsettling. To Bob Barr, writing in the *Michigan Law Review* [2], it is "a chilling vision" and R. S. Deese, in *We Are Amphibians* [3], calls its premise "the mass production of human beings."

The discovery in 1953 of the structure of DNA led to the advent of real genetic science that could change people. DNA editing appears in several films analyzed by the film historians David A. Kirby and Laura A. Gaither in "Genetic Coming of Age: Genomics, Enhancement, and Identity in Film" [4]. The authors single out *Gattaca* as showing a society that has "so much confidence in the predictive power of genomics that their culture revolves around these expectations." The film provides a lesson in the eugenic effects of editing human DNA. Its title combines the first letters of guanine, adenine, thymine, and cytosine, the "base pair" compounds essential to how DNA transmits genetic information.

The social order in *Gattaca*, set in the "not-too-distant future," is far looser than in *Brave New World*. It is much like today's world with one crucial change: Genetic science has advanced so that a person's genetic makeup can be easily tested, and it is routine to alter the DNA of an embryo to produce a baby with specified characteristics. The result is a society dominated by "genetic destiny."

Genetic augmentation is not available to everyone in this society. Only those with means can pay geneticists to implant assets like good looks or musical ability in the DNA of their children-to-be. Although it is illegal to discriminate on the basis of a person's genetic profile, in practice "Valids," those with superior genetic credentials, have every advantage and live desirable lives, whereas the less genetically favored "In-valids" or "De-gene-rates" are the Epsilons of this society, who push brooms and clean toilets.

In the story, young Vincent Freeman (Ethan Hawke) is a non-augmented In-valid who is projected to develop serious medical conditions. Through sheer grit and refusal to quit, he physically outperforms his enhanced Valid brother, determined to realize his ambition of becoming an astronaut. The closest he can come, however, is to work as a janitor at the Gattaca Aerospace Corporation, which launches space missions.

Vincent games the system by acquiring the superb DNA profile of Jerome Morrow (Jude Law), a former Olympics swimmer now in a wheelchair because of an accident. After surgery to make himself resemble Jerome (and with Jerome's help), Vincent can pass as a Valid. His passion affects the disabled Jerome, who famously declares: "I only lent you my body. You lent me your dream." Now apparently genetically qualified, Vincent is selected for astronaut training. In the final scene, we see him blast off on a mission to Titan, one of Saturn's moons.

Any science that professes to predictably change humanity should be carefully weighed—or its results may come to haunt us and the new humans we make. *Brave New World* shows an extremely repressive society whose eugenic system keeps a select group in control. Although such a goal might appeal to the far right, in the near term, at least, it is hard to imagine such a movement gaining the political power to impose a Nazi-like program of gene editing.

Gattaca, however, presents a believable model for the future. It reflects and extends current attitudes toward race and the disabled, and with America's growing gap between haves and have-nots, its speculations ring true. Buying genetic advantage to give one's child an edge in life would be just a step beyond what parents now do—sending a very young child to an expensive private school, for instance—to gain that edge.

In a United States where medical care is not equally available to all, genetic enhancement will likely be too costly for all but the wealthy. As in *Gattaca*, buying enhancement will not be illegal, nor seen as unethical. But it would widen existing health and social inequalities, as expressed in the reactions to the "Human Genome Editing" report. Those who can afford it would choose mental and physical advantages for their offspring, perhaps including traits such as selfishness or "win at all costs" personalities that might benefit them but harm society. This would enhance a special group that would not need Francis Galton's selective breeding to make itself superior over time, leaving everyone else as the In-valids.

This approach could also erode America's racial and ethnic diversity, fulfilling Rep. King's fantasies. Homogeneity is exactly what would result if a favored group genetically replicates and enhances itself to produce future generations with the same appearance and attitudes, only more so.

In the final analysis, *Brave New World* portrays a "hard eugenics" created by a government to suppress human rights, diversity, and opportunities for its citizens. But like the world in *Gattaca*, our own society could instead display a eugenic element not imposed from above, but arising from our society's dynamics. Unless our society balances the undoubted benefits of gene editing against its equally undoubted risks, the greater danger may come not from authoritarian government but from this "soft eugenics."

References

1. Lennox, William. "Should they live? Certain economic aspects of medicine," *The American Scholar* **7**, 4 (1938), pp. 454–466

2. Barr, Bob. "Aldous Huxley's *Brave New World*—Still a chilling vision after all these years," *Michigan Law Review* **108**, 6 (2010), pp. 847–857

3. Deese, R. S. "Twilight of Utopias," in *We Are Amphibians: Julian and Aldous Huxley on the Future of Our Species*, pp. 56–85, University of California Press (2015).

4. Kirby, David A. and Gaither, Laura A. "Genetic coming of age: Genomics, enhancement, and identity in film," *New Literary History* **36**, 2, Essays Probing the Boundaries of the Human in Science (2005), pp. 263–282.

The Case Against an Autonomous Military

In 2016, a Mercedes-Benz executive was quoted as saying that the company's self-driving autos would put the safety of its own occupants first. This comment brought harsh reactions about luxury cars mowing down innocent bystanders until the company walked back the original statement. Yet protecting the driver at any cost is what drivers want: A recently published study in *Science* (available to read on *arXiv*) shows that, though in principle people want intelligent cars to save as many lives as possible (like avoiding hitting a crowd of children, for example), they also want a car that will protect its occupants first.

It would be hard to trust this algorithm because we—the humans nominally in charge of the AI—don't ourselves have the "right" ethical answer to this dilemma. Besides, and potentially worse, the algorithm itself might change. No matter how a car's AI is initially programmed, if it is designed to learn and improve itself as it drives, it may act unpredictably in a complicated accident, perhaps in ways that would satisfy nobody. It would be unsettling as a driver to worry whether your car really wants to protect you; fortunately, ethically ambiguous driving situations do not happen often—and those might happen even less if self-driving cars fulfill their touted promise of high competence, hugely reducing the frequency of auto accidents, though this remains to be seen. In March 2018, a Tesla Model X, engaged in Autopilot, slammed into a California highway median and caught fire, killing the driver (who was warned several times by the car to place his hands on the wheel), and, in Arizona, a self-driving Uber fatally hit a pedestrian.

The potential harm of AIs deliberately designed to kill in warfare is much more pressing. The United States and other countries are working hard to develop military AI, in the form of automated weapons, that enhance battlefield capabilities while exposing fewer soldiers to injury or death. For the United States, this would be a natural extension of the existing imperfect drone warfare program—failures in military intelligence have led to the mistaken killing of non-combatants in Iraq. The Pentagon says that it has no plans to remove humans from the decision process that approves the use of lethal force, but AI technology is out-performing humans in a growing number of domains so fast that many fear a run-away global arms race that could easily accelerate toward completely autonomous weaponry—autonomous, but not necessarily with good judgment.

Thousands of employees at Google are worried about this outcome. Google has been lending its AI expertise to the Pentagon on its Project Maven, which has been working since 2017 to establish an "Algorithmic Warfare Cross-Functional Team" that will eventually allow one human analyst, according to Marine Corps Colonel Drew Cukor, "to do twice as much work, potentially three times as much, as they're doing now" on tasks like recognizing targets—road-side bomb planters, for instance—in drone video footage. "In early December, just over six months from the start of the project," according to the *Bulletin of the Atomic Scientists*, "Maven's first algorithms were fielded to defense intelligence analysts to support real drone missions in the fight against ISIS." In a petition signed by over 3,100 employees, including many senior engineers, according to the *New York Times*, the employees asked that Google stop assisting Project Maven, stating, "We believe Google should not be in the business of war."

Even if Google pulls out, other companies developing AI, like Microsoft and Amazon, will likely step in—Amazon is thought to be the favorite to win the Pentagon's multibillion dollar cloud services contract, which includes AI technology. So, it's all too easy to imagine—in the not-so-far future—one of these companies helping the military design an AI that chooses targets, aims, and fires with more speed and accuracy than any human soldier, but lacks the wider intelligence to decide if the target is appropriate. This is why the U. N. is considering how to place international limits on lethal autonomous weapons. In 2015, thousands of researchers in AI and robotics asked that autonomous weapons be banned if they lack human control because, among other reasons, "autonomous weapons are ideal for tasks such as assassinations, destabilizing nations, subduing populations, and selectively killing a particular ethnic group."

These terrible disruptions would be within the grasp of specialized AI capabilities that exist now or will in the near future, such as facial recognition, making it possible to hunt down a particular target for assassination; or worse, algorithms that pick entire groups of human targets solely by external physical characteristics, like skin color. Add to that a clever dictator who tells his people, "My new AI weapons will kill them but not us," and you have a potential high-tech crime against humanity that could come more cheaply than building rogue nuclear weapons. It's certainly not far-fetched to suppose that a group like ISIS, for example, could acquire AI capabilities. Its supply chain of weapons—some of which they've creatively repurposed for urban combat—has been traced back to the United States. "For ISIS to produce such sophisticated weapons marks a significant escalation of its ambition and ability," *Wired* reported in December. "It also provides a disturbing glimpse of the future of warfare, where dark-web file sharing and 3-D printing mean that any group, anywhere, could start a homegrown arms industry of its own"—a tempting possibility for international bad actors to expand their impact.

We already have enough problems dealing with the large-scale weapons capabilities of nations like North Korea. Let's hope we can keep AI out of any arsenal developed by smaller groups with evil intentions.

Frankenstein Turns 200 and Becomes Required Reading for Scientists

Review: *Frankenstein: Annotated for Scientists, Engineers, and Creators of All Kinds.* Mary Shelley, with David H. Guston, Jason Scott Robert, and Ed Finn, editors (MIT Press, 2017).

Everything about Mary Shelley's *Frankenstein* (1818) is remarkable: not just the story about the laboratory creation of a living being, but its backstory. Mary started writing it in 1816 as an 18-year-old. The daughter of two eminent English intellectuals, she ran away to France at 16 with the married romantic poet Percy Bysshe Shelley. She lost their child soon after its birth, and then married Percy after his wife committed suicide. The story was written at the instigation of Percy's literary pal, the famously "mad, bad, and dangerous to know" Lord Byron, who had his own romantic entanglement with Mary's stepsister. When the book emerged, critics first ascribed it to Percy as a story that could not or should not have been written by a woman, until Mary corrected the record in her revised edition of 1831.

This rocky beginning turned into one of the greatest literary coups of all time. The tale Mary conjured up has never been out of print since it was published, and it is said to have appeared in more editions than any other novel. It has inspired scores of adaptations and related works in film, theater, and television. Now in its 200th anniversary year, it is the subject of a great many analytical and celebratory books—among them, *Frankenstein: Annotated for Scientists, Engineers and Creators of All Kinds.*

This effort was funded by the National Science Foundation as a "science and society" project. Its publication by MIT Press is important for what it says about the current scientific mood. It seems clear that the separation of what C. P. Snow in 1959 called "the two cultures" is no longer tenable if our species (and the planet) is to prosper, let alone survive. Humanists have long said that science needs the humanities. Now scientists themselves and the scientific establishment seem to be on board, acknowledging that we need to read and creatively imagine "what if" scenarios lest we wear blinders. A significant indicator of this new mood: the second issue this year of the internationally influential research journal *Science* featured a cover image, an editorial and a long article devoted to "the lasting legacy of Frankenstein." An 18-year-old girl's literary creation is now required reading, as it were, for scientists.

In the same spirit, the three editors of *Frankenstein: Annotated*— David H. Guston, Ed Finn, and Jason Scott Robert—describe their endeavor as nothing less than harnessing the "galvanizing power of *Frankenstein*" to "ignite new conversations about creativity and responsibility among science and technology researchers, students, and the public." The editors, who hold faculty positions at Arizona State University (ASU), hail from disciplines ranging from science and technology to science policy and the humanities and ethics. (Full disclosure: I attended a related *Frankenstein* workshop Guston and Finn ran at ASU in 2014, which motivated me to create, co-edit, and contribute to my own *Frankenstein* book this year. I myself am a scientist, not a humanist.)

Their expertise speaks to *Frankenstein*'s enduring message about existential stakes—and the potentially alarming societal consequences likely to devolve from the unfettered march of science and technology. Concerns about unintended consequences were urgent at the onset of the Industrial Revolution and the Nuclear Age, and they are, if anything, more urgent now. As the editors write in their introduction, we are now on the verge of hugely consequential technological endeavors:

> the creation and design of living organisms through techniques of synthetic biology [and] of planetary-scale systems through climate engineering, and the integration of computational power [into] global society...each presents real and even existential risks...Who gets to decide on the agenda for scientific research... Who gets to say what problems or grand challenges we try to solve? Who gets to say how we solve them... Who gets to partake in those benefits...

In the spirit of public education, they use different elements or strategies to enhance what they call "our collective understandings." These include, for instance, annotations to the text of the 1818 *Frankenstein*; Mary's introduction to the 1831 edition, which relates how the story came to be; short essays by contributors of various stripes; and appendices including "[d]iscussion questions" appropriate for students and others.

All this is introduced in a long piece about Mary and her work by the late literary scholar Charles E. Robinson of the University of Delaware, who had earlier extensively analyzed Mary's original manuscript. He describes how the scientific knowledge and attitudes of the time helped Mary imagine that a living being might be constructed. Mary was not a Luddite, he writes, and he surmises her parents encouraged her scientific interests. Both were writers and political philosophers, with Mary Wollstonecraft known for her book *A Vindication of the Rights of Woman* (1792) and William Godwin known for his radical political views.

Though Wollstonecraft died soon after Mary's birth, her feminist writings espoused equal education for boys and girls in "natural philosophy." Mary knew of the theories of Charles Darwin's grandfather Erasmus Darwin about the origins of life; she likely met her father's scientific friend William Nicholson, co-discoverer of electrolysis; and she read Humphry Davy's book about chemistry while writing *Frankenstein*. Percy, also interested in science, had carried out electrical experiments when he was at Oxford, and the couple attended lectures in London about chemistry and electricity. After Luigi Galvani's research in "animal electricity" in the 1770s, electricity was considered a possible "spark of life."

Robinson relates these experiences to characters and events in *Frankenstein*, then adds:

> *Frankenstein* and this introduction encourage STEM students to respect the humanities as offering a valid means of defining and even improving the world, much as science hopes to do. *Frankenstein*...has become a metaphor for science that ignores human consequences and values.

In a nod to the students the book hopes to reach, Robinson ends by showing how *Frankenstein* has been reimagined in more recent media presentations. In optimistic Hollywood versions, artificial beings are shown valuing humanity and saving lives—for instance, in James Cameron's film *Terminator 2: Judgment Day* (1991), when the T-800 android (Arnold Schwarzenegger) comes from the future to help avert the nuclear exchange programmed to destroy humanity. Of course, there are plenty of less optimistic spin-offs.

But Robinson does not dwell on these. His introduction is followed by the text of the 1818 *Frankenstein* with well over a hundred brief annotations crowd-sourced from nearly 50 contributors—academics in the humanities, natural and social sciences, and independent scholars, writers, and artists. This approach provides some fresh takes on *Frankenstein*—for instance, a note from an Earth scientist links the passion for scientific research in the early 19th century to the search for land and resources beyond the frontier. Many of the notes though, about literary and historical allusions, the motivations of the characters, the science of the time, and scientific ethics, will not surprise anyone familiar with *Frankenstein* and its milieu. But they are useful for students and first-time readers of the story.

Seven short formal essays follow the annotated text. In the first essay, Josephine Johnston, the director of research at the Hastings Center, a bioethics institute, defines "responsibility" as either "a *duty* to take care of something or someone or the *state* of being the cause of an outcome." The monster's creator, Victor Frankenstein, she writes, shirks both forms of responsibility. He does not think through the consequences of his work, and in the creation scene in the novel, he rejects the unnatural being he has made immediately after it stirs to life.

Frankenstein draws attention to this quasi-parental failure when the monster complains to Victor,

> But where were my friends and relations? No father had watched my infant days, no mother had blessed me with smiles and caresses [...] I had never yet seen a being resembling me, or who claimed any intercourse with me. What was I?

and begs him to fashion a female companion. The monster's loneliness is palpable. Unlike climate engineers who "merely" create technological disruption, the creators of life are also responsible to the sentient, feeling beings they make. If their aim is to make beings that are not monstrous, but accepted and content within human society, then this adds another ethical layer to their endeavor.

In another essay, Jane Maienschein, director of the Center for Biology and Society, and Kate MacCord, a PhD candidate in the School of Life Sciences, both at ASU, emphasize the importance of human development. Titled "Changing Conceptions of Human Nature," their essay retraces Aristotle's thinking about deviations from the normal characteristics of a species, which the authors interpret as a path to the rise of a "monster." Aristotle argued that individuals must develop over time to become fully human, much as modern biology now confirms. In animating a physically complete being, then abandoning it,

> Victor made the fatal mistake of failing to understand that producing a life, in the sense of a fully and properly functioning living human, requires development [...] It is not clear whether Victor or Mary learned the lesson that development matters or fell for the illusion that matter is enough.

Others agree that development matters in creating artificial life. In his seminal 1950 paper "Computing Machinery and Intelligence," Alan Turing, author of the eponymous test for artificial intelligence, asked "Can machines think?" The proper approach to building an AI, he believed, is to construct one at a child's level, then educate it up to an adult level, which he thought would take as long as educating a human child.

In *Frankenstein, Gender, and Mother Nature*, Anne K. Mellor, who works in English literature and women's studies at UCLA, takes a different tack. The novel, Mellor writes, "brilliantly explores what happens when a man attempts to have a baby without a woman [and why] an abandoned and unloved creature becomes a monster." As other critics have done, she relates these elements to Mary's feelings of "isolation and abandonment" after her mother's death and her father's remarriage to a stepmother who treats her badly, and to Mary's fears about her own pregnancy.

Calling *Frankenstein* a feminist novel, Mellor argues that Victor reveals his fears of female sexuality when he destroys the female counterpart he had started to build for his creature. He is creating a men-only society. The feminist elements in the story are clearly still meaningful—for instance, in the recently released biopic *Mary Shelley* by the pioneering female Saudi Arabian director Haifaa al-Mansour. She recognized the similarities between Mary's struggle to establish herself as a writer and her own background "as an aspiring artist in a conservative Muslim culture," and so directed a film described as "suffused with a righteous feminist fire."

Inevitably, with so much already written about *Frankenstein*, many of the essays echo themes that have been explored elsewhere. But "The Bitter Aftertaste of Technical Sweetness" by Heather Douglas stands out in a book devoted to helping scientists think about the implications of their work. Douglas, a philosopher of science at the University of Waterloo, Canada, relates a historical case study, the World War II Manhattan Project, to how scientists really feel about their research and how that can affect ethical considerations. She describes "technical sweetness" as the moment when "all the pieces [of a scientific puzzle] fit beautifully and functionally together." The sweetness "is alluring, consuming and, as we can see in the story of Victor Frankenstein, potentially blinding to what might follow from the solution being sought."

The Manhattan Project scientists overcame huge obstacles to build an extremely powerful but controllable atomic bomb. They exulted when a first test at Alamogordo succeeded on July 16, 1945, but they were plagued by nagging doubts even before the destruction at Hiroshima and Nagasaki. "Now we are all sons-of-bitches," said one scientist. Another resigned from the project but was not allowed to share his reservations with his colleagues. Later, others recognized their responsibility in opposing ways. Some supported international control of nuclear weapons, whereas others advocated for the far more destructive hydrogen bomb to counter the Soviet Union. In other words: Even when we accept responsibility, what ought to come next is rarely clear.

Douglas notes that the Manhattan Project differed from Victor's solo effort in important ways: it employed thousands of scientists and was driven by real or perceived wartime needs and fears. But both projects functioned in isolation, separate from societal values. The Manhattan Project was kept secret, and as Douglas writes, Victor stopped "communicating with his friends and family and [disengaged] from the social connections that might give him a better perspective on his pursuit." And in both cases, the impetus toward technical sweetness—closely related to the "technological imperative," the desire to do it "because we can"—delayed recognizing consequences.

In these and other ways, *Frankenstein* offers lessons in how research today could go wrong, though to be sure much of the scientific community does indeed now recognize the need for ethical brakes—to control the modern equivalent of what Victor attempted, namely genetic manipulation using synthetic biology and the new CRISPR gene-editing technique.

In 2015, Nobel Laureate David Baltimore and other scientists proposed placing a moratorium on research that would cause heritable changes to the human genome. In 2016, when the geneticist George Church at Harvard and a few colleagues announced their intention to construct the entire human genome from scratch, concerns about the lack of broader input forced the group to scale back its ambitious agenda. And in late 2017, the National Academy of Sciences and the National Academy of Medicine published *Human Genome Editing: Science, Ethics, and Governance*, a guideline for ethical genetic research that addresses the specific questions raised by the editors of this annotated *Frankenstein*.

But there are no guarantees that these standards will be followed. Moreover, though research in genetic manipulation is not driven by wartime considerations as was the case with the Manhattan Project, it is driven by commercial ones. At least a billion dollars in start-up funds has been funneled into companies that together form what has been called CRISPR Incorporated. These companies plan to apply genetic techniques to medicine and agriculture for profit—a powerful incentive to deliver the technology rapidly and perhaps with minimal forethought.

Frankenstein: Annotated for Scientists, Engineers, and Creators of All Kinds can remind scientists and engineers to proceed with caution. Though not everything in the book works equally well, it offers a valuable set of approaches to the ethical questions the original *Frankenstein* raises. By itself or in the hands of a teacher, *Frankenstein: Annotated* could encourage practicing or future scientists and technologists to consider the impact of their work, and it might remind ordinary citizens that science affects their lives when it leaves the lab.

Instilling these attitudes in young people could help change the world. We also need a gifted high school or college student of Mary's age to write his or her own *Frankenstein* for the 21st century. If this annotated edition inspires even a single young person to write such a book, then *Frankenstein: Annotated* will have been a success.

Can a Physics of Panic Explain the Motions of the Crowd?

When people come together in a crowd, physical and emotional connections define their movement, state of mind and will to act. Understanding crowds can help us manage the panic caused by a terrorist attack; a science of crowds is vital to managing many emergencies, especially when density becomes dangerously high. Panic or chaos in a crowd can kill or injure hundreds, as happened at the Love Parade in Germany in 2010 when thousands of attendees to an electronic dance music festival piled up as they tried to enter a narrow tunnel; 21 people died of suffocation.

Fundamental science and public safety demand that we develop a complete science of crowds using a range of disciplines. Today, work by social psychologists shows that crowds are influenced by the personalities of individual members; thus, crowds can embody altruistic and helpful behavior as well as the opposite. And now we can extend crowd science further by incorporating quantitative analysis using classical and statistical physics, computational science, and the theory of complex systems—the study of groups of interacting entities.

One relevant concept from complexity theory is "emergence," which occurs when the interactions among the entities produce group behavior that could not have been predicted from the properties of any individual element. For instance, randomly moving H_2O molecules in liquid water suddenly link up at zero degrees Celsius to make solid ice; starlings in flight quickly form themselves into an ordered flock.

Emergent behavior can be predicted if the interaction among the entities is known, as shown in 2014 by researchers at the University of Minnesota who determined how two people in motion interact and, from that, how a crowd moves. The researchers first considered an idea from physics, theorizing that, like electrons, pedestrians avoid collision by repelling each other as they get closer. But video databases showed instead that when people see that they are about to collide, they change their paths. From this, the researchers derived an equation for what amounts to a universal force of repulsion between two people, based on time until collision, not distance.

The formula successfully reproduced the emergent real-world features of a crowd, such as forming a semicircular configuration while waiting to trickle through a narrow passage, or extemporaneously developing independent lanes as its members walk toward different exits. This makes it possible to simulate crowd behavior to design evacuation routes, for instance.

To be useful in emergencies, crowd analysis must also account for emotional contagion. Spreading fear can change emergent behavior, as shown by researchers at the K. N. Toosi University of Technology in Iran. In 2015, they created a computer version of a public space populated with hundreds of simulated adults and children, and security guards who directed people to the exits. Assuming that the participants were responding to a dangerous event, the simulation escalated them to greater levels of fear and panicked, random movement when they failed to find an exit.

Running the simulation, the researchers found that between 18 and 99 per cent could escape, depending on the combination of participants. The greatest number of escapes did not occur with the smallest or largest numbers of people or security agents but at intermediate values. This shows that the emotional state of a crowd can carry its dynamics into a complicated non-linear stage.

We can determine the emotion of individuals in a real crowd by observing their physical behavior. In 2018, a team under Hui Yu of the University of Portsmouth in the U. K. used kinetic energy, the energy of motion in physics, to serve as a gauge that could establish when a crowd enters an "abnormal" emotional state. Crowd members running from a dangerous event such as an explosion have increased kinetic energy, which can be detected in real-time crowd video images. Using computer vision techniques, the researchers calculated the speeds of the pixels that make up the images, from which they identified the most energetic part of the crowd.

The researchers applied their method to the dataset of video clips collated by the computer scientist Nikolaos Papanikolopoulos and colleagues at the University of Minnesota. Those clips show crowds of real people reacting to simulated emergencies. Initially, the subjects walk normally, then suddenly disperse and run in all directions. The energy algorithm quickly sensed these transitions, and the researchers conclude that the method can automatically detect unusual, potentially dangerous behavior in public gatherings.

Other links between emotions and actions have been drawn by the computer scientist Dinesh Manocha at the University of Maryland and his colleagues in their "CubeP" model, which unites analysis of factors from physics, physiology, and psychology. These three factors are strongly interrelated during the physical activity and emotional responses that mark a crowd in crisis. CubeP uses the basic physics of forces and velocities to calculate the bodily effort of a person in motion. CubeP also incorporates the model of emotional contagion developed in 2015 by the computer engineer Funda Durupinar at Bilkent University in Turkey and her colleagues, which includes typical personality profiles that determine a person's response to stress. CubeP adds a physiological measure of the panic level for each person, based on bodily effort. This affects heart rate, which is known to indicate the degree of fear. All this is combined to predict the speed and direction of motion for each crowd member.

The researchers tested CubeP in computer simulations of a crowd reacting to a dangerous event, with realistic results. A virtual person near the threat quickly panics and runs. A more distant individual responds to emotional contagion with fear and escape behavior, though later. The researchers also applied CubeP to the University of Minnesota dataset and to videos of real emergencies, such as on the Shanghai subway system in 2014, and outside the British Parliament building in 2017. In all these, CubeP simulations of crowd behavior were reasonably close to reality, and closer than the Durupinar approach and other models that do not merge physical, psychological, and physiological factors.

This improvement illustrates the power of a multidisciplinary science of crowds. As the insights accumulate, they are sure to be useful in architectural design and disaster planning. Findings could, however, lead to more surveillance of crowds in public spaces, a phenomenon currently raising concerns from the American Civil Liberties Union about privacy and potential for abuse.

Something is lost and something gained by reducing crowd behavior to numbers. Comparing models to real data will provide welcome insights into crowd dynamics, but we need a sweeping understanding from psychology as well. Elias Canetti, the Nobel Prize–winning author who wrote the classic *Crowds and Power* (1960), foresaw the day when this partnership would help to break the crowd code. In considering the importance of a certain critical density in crowd behavior, he wrote: "One day it may be possible to determine this density more accurately and even to measure it." Now we can measure and analyze such quantities, but we also need the expansive views of the humanities and social sciences to tell us what they really mean.

Fantasy into Science: Invisibility

Fantasy fiction is about magic, science fiction is about...well, science. People who believe in one don't always buy into the other, yet the two can merge; as Arthur C. Clarke wrote, a sufficiently advanced technology can't be distinguished from magic. Besides, some magical visions represent such deep human yearnings that we ardently wish they were real.

These visions often appear in fantasy, myth, and legend, where they're "explained" simply as being magical. They might reappear in the newer genre of science fiction, to be newly justified by more or less credible scientific explanations. And sometimes, if we're lucky, there's a third step where the idea moves out of fiction altogether and becomes real.

If you think this doesn't truly happen, consider the common human dream of flying. The vision of freely soaring through space has been expressed in the old Greek story about Icarus, who flew with wings made of feathers and wax; in the flying carpet of the *Arabian Nights*; in Jules Verne's *Around the World in 80 Days*, where an aeronaut circles the globe by balloon; and now, we routinely fly at 600 miles per hour. "Flight" has moved from myth and fairy tale to scientific possibility to everyday reality.

But the dream of becoming invisible has seemed only an impossible fantasy for 2,500 years, from Plato's *Republic* to *The Lord of the Rings* and the Harry Potter stories. It became a science-fiction theme in tales like H. G. Wells' *The Invisible Man* (1897), which described an optical approach to it, and in film and TV. Even with limited special effects, the 1933 film version of *The Invisible Man* entertainingly displays personal invisibility. Later, *Memoirs of an Invisible Man* (1992) portrays Chevy Chase in a "state of molecular flux" that makes him invisible. In *Hollow Man* (2000), spectacular computer-generated imagery (CGI) shows how scientist Kevin Bacon turns invisible after his biomolecules are tweaked. And beyond personal invisibility, *Star Trek* gave the hostile Romulans a "cloaking device," based on Einstein's general relativity, that hides an entire spacecraft.

Now, after roles in fantasy and science fiction, invisibility has become real. One method, stealth technology, has since the 1980's made U. S. warplanes nearly invisible to radar. The aircraft is coated with absorbing material to reduce its overall reflectivity, and more important, is shaped so radar beams reflecting off it are diverted in directions where they won't be detected. In the 1950's, the B-52 Stratofortress bomber was highly visible on radar screens with a huge cross-sectional area of 150 square meters. Today, stealth aircraft like the F-22 Raptor fighter have radar cross-sections the size of an insect, and the technology is still developing, as in a stealth helicopter that participated in the recent U. S. raid on Osama bin Laden's compound in Pakistan.

A second method, invented in 2003 by Susumi Tachi at the University of Tokyo, makes an object *apparently* disappear. The front of the object is coated with "retroreflective" material, which sends incoming light directly back out. A video camera records what lies behind the object and this image is projected onto the object's retroreflective front, so that an observer sees that scene superimposed on the object. The result is a remarkable illusion, as shown in videos where crowded street scenes are seemingly visible right through people wearing retroreflective cloaks. This isn't yet Harry Potter's Cloak of Invisibility; the optics must be carefully set up and the method works only for a static observer in the correct position. Still, it has applications in military camouflage and in enhanced visibility from automobiles and aircraft.

Newest and most striking, however, are methods that could make things truly disappear as completely as in any fantasy. We see an object as it interacts with light rays, for instance, by changing their direction. But in principle, a cloak placed around an object could intercept incoming light rays, bend or refract them into itself, and send them along internal paths so that they leave the cloak along continuations of their original trajectories. An observer would see only apparently undisturbed light and so would conclude there's nothing there—much like water in a stream encountering a rock, splitting around it, then smoothly recombining to give no sign downstream that the rock exists.

In 2006, two separate research groups applied breakthroughs in understanding how light interacts with matter to work out the mathematical theory of cloaking. The same year, a group under D. R. Smith at Duke University used this approach to build the first true cloak, an intricate artificial material or "metamaterial." Its thousands of millimeter-size copper units were carefully shaped to change the electric and magnetic properties of light—an electromagnetic wave—as it passes through, to bend the light along the desired paths. Tested at microwave wavelengths, the cloak made a copper cylinder nearly completely disappear.

This was huge, exciting news, but to make a Harry Potter cloak, the method needs to be extended to visible light. Since the units in a metamaterial must be smaller than the wavelength of the light, it takes advanced nanotechnology to make a cloak for the tiny visible wavelengths 400 to 750 nanometers, violet to red. Still, scientists are successfully building appropriate structures, and there are other encouraging results. Using the unusual optical behavior of the naturally occurring crystal calcite, researchers have created "carpet" cloaks that make a bump appear flat. This hides small objects underneath, and works for red, green, and blue light.

If your dream is to actually wrap a cloak around yourself, that too could be on the horizon. In 2002, Italian scientists proposed that a flexible cloak covered with many small light sensors and emitters could do what the retroflection method does—records what lies in back and projects that image in front, effectively making the cloak disappear—but dynamically, in real time, to maintain the effect as the cloak or an observer moves. Though we're not there yet, we have candidates for the necessary devices; quantum dots, tiny bits of semiconductor that behave like artificial atoms. These can both detect and produce light, and can even be embedded in a cloak by spraying them onto cloth via inkjet printer.

Just five years after that first report of cloaking at Duke, it has been referenced by other scientists over 800 times, showing how much effort is going into the practical realization of invisibility. It's easy to imagine the wondrous, centuries-old power of invisibility turning into a new consumer product like "InvisiWrap™—what happens under the Wrap stays under the Wrap!;" scientific magic, or magical science, at the bargain price of $49.95.

Fantasy into Science: Teleportation

Over many years, we have watched Captains Kirk and Picard, and others in the *Star Trek* universe step onto a transporter platform, fade into shimmering motes of light, then instantaneously reappear on the surface of an unexplored planet. This is teleportation, surely the coolest possible way to travel through space—no need to squirm into a spacesuit or navigate a spaceship; just step up and away you go!

Except that in the real world, away nobody goes, because teleportation has just been pure fantasy. Long before the term was invented in 1931, instantaneous or at least extremely rapid travel was a magical favorite. In European folklore, seven-league boots enable their wearer to cover that much ground, 21 miles or nearly 35 kilometers, per step. In the tale of Aladdin and the lamp, a jinn instantly transports himself and an entire palace from China to Morocco. Likewise, Siegfried in Wagner's Ring cycle travels long distances in an eye-blink via the Tarnhelm, a magical helmet.

However, in dozens of science-fiction scenarios, not just *Star Trek*, people are dematerialized, beamed elsewhere, and reassembled not by magic but by technology. Still, there is the danger that the pattern representing you might become distorted as it is transmitted, or worse yet, become confused with someone or something else. That is the awful fate of the researcher in the film *The Fly* (1958, with Vincent Price; 1986, with Jeff Goldblum), who has the bad luck to start teleporting just as a random fly buzzes into the chamber. What emerges is a repulsive combination that neither a human Mom nor a mother fly could love.

Though we do not yet need to worry about this grisly outcome, astonishingly teleportation is now known to truly work at least at the level of elementary particles. It arises from the strange but real quantum effect of "entanglement," in which two submicroscopic particles emerging from a process at the same time remain forever linked thereafter, no matter how widely separated. That is, measuring the properties of one immediately determines the state of the other as if they are somehow communicating by some unknown means, even if many kilometers apart.

To see where the weirdness enters, think of two electrons, A and B. Electrons behave like tiny bar magnets and two of them can be correlated to have net zero magnetic field so that whatever direction the North pole of A points, B points oppositely, cancelling it out. So far, so good; but according to entanglement if you measure the North pole of A as pointing straight up, electron B, no matter how far away, responds by pointing its North pole down; and if you measure A as pointing down, B would be found pointing up. For a simple analogy, imagine pairs of socks correlated to always have the same color; then entanglement predicts that if you randomly pick a sock from a pile of black and white ones, a second sock randomly chosen from a different pile always matches the original color. But how does the second sock "know" the color of the first sock? How does electron B know, and respond to, the state of electron A?

No one understands what lies behind this apparent violation of ordinary space and time. It deeply troubled Einstein, who called it "spooky action at a distance;" but it is absolutely confirmed by experiments using pairs of photons, atoms, and electrons separated by up to dozens of kilometers. Equally perplexing, measurements show that any communication between the two members of the pair travel at multiples of the speed of light and maybe even instantaneously. This would upset Einstein as well, because his theory of relativity establishes that nothing can go faster than light.

Nevertheless, in 1993, physicist Charles Bennett at IBM and his colleagues embraced the weirdness to theorize that entanglement could enable what they aptly called "teleportation" as a "term from science fiction meaning to make a person or object disappear while an exact replica appears somewhere else." Instead of electrons, they considered photons, the quantum particles of light. These carry electric fields that can be pointed in specific directions, or "polarized," and polarized photons can be made in correlated pairs. Bennett's group showed that the link between entangled A and B photons could transmit the polarization of a third "X" photon located near A to a distant B photon, turning it into a perfect copy of X, which itself disappears in the process—the very definition of teleportation.

A photon was actually teleported for the first time in 1997, followed by other experiments that confirm the process for pairs of photons and also of atoms separated by several kilometers.

Now the mysteries of entanglement and teleportation contribute to technology. They can be used to send photons through optical fiber telecommunications lines to represent digital data in an absolutely unbreakable code. This ability is already being used to securely transfer financial information. You might soon encounter teleportation technology that guarantees the security of messages from your smartphone to your bank or your Facebook friends. Down the road, researchers foresee powerful new computers based on these quantum principles.

Still, these applications are only microscopic versions of beaming Captain Picard through space. We may never be able to teleport people or big objects because it is theoretically impossible to transmit the necessary huge mass of information without distortion. The last thing anyone wants is hi-tech quantum teleportation that reproduces the hybrid horror of *The Fly*; but then, if banks were to teleport photons to protect their customer's financial data, you might be lucky enough to have your account mistakenly merged with that of some random billionaire.

Fantasy into Science: Tractor Beams

Short of destroying a whole world with planet-breaking weapons, the most action-filled moments in science fiction come when opposing spacecraft clash. As phasers fire and missiles launch, ships frantically maneuver, attack, or spectacularly explode. But sometimes the aim is to capture a spaceship intact, or if she has superior speed, to grab and hold her while battering down her defenses. That is the space version of a fighting technique from the days of wooden sailing ships, which is to pull an enemy ship close and hold her with grappling hooks and ropes, then board her, or pound her into wreckage at point-blank range.

In space, this tactic needs a futuristic version of hooks and ropes—a tractor beam. Like the force of gravity in nature, a tractor beam pulls things toward its source; but in science fiction, it is stronger than ordinary gravity, and unlike gravity, it can be aimed. Tractor beams first showed up in the Buck Rogers stories and the "space operas" of Edward E. "Doc" Smith from the 1920s and 1930s. They still appear in *Star Trek*, the film *District 9* (2009), and other contemporary science fiction.

Gravity may feel all too strong when you are jogging uphill, but technically it is the weakest of the universe's four fundamental forces (in descending strength, the others are the strong nuclear, electromagnetic, and weak nuclear forces). Gravity's pull takes on real power only when it arises from a massive object like a planet or a star. The mass of even the biggest spacecraft could not generate enough force to yank in another spacecraft. Going beyond this to create a powerful, directed artificial gravity needs a technology we do not yet have a clue about.

But there is another way to manipulate objects without touching them, using something you would not think has physical impact: light. Light's intangibility might seem to imply that it cannot affect solid objects. By the weird rules of quantum mechanics, however, light is both an electromagnetic wave and a flock of particles—photons—that carry energy and momentum. As Isaac Newton worked out long ago, a change in momentum produces a force; and so, no differently from baseballs or billiard balls, when photons hit something and thus change their momentum, they exert a small force called radiation pressure.

On Earth, the radiation pressure from sunlight is 50 million times weaker than atmospheric pressure, but a laser can intensify the effect to a useful level. In the optical tweezers method, developed at Bell Laboratories in 1986, a focused laser beam suspends small biological objects like bacteria and DNA molecules in mid-air and can be used to manipulate them. On a larger scale, the Sun's radiation pressure can drive a spacecraft, if the ship is outfitted with a big sail that can capture the push from many photons. In 2010, the IKAROS spacecraft launched by JAXA (Japan Aerospace Exploration Agency) deployed a sail with an area of 200 square meters and headed off toward the planet Venus, accelerated by sunlight.

These applications may not be all that surprising, because it is easy to picture tiny bullets of light pushing an object. But could light possibly attract rather than repel a body, and would that be of any earthly use? To NASA, the answer to both questions is "yes." The agency envisions "unearthly" uses such as cleaning up the orbital debris ringing our planet and pulling an incoming space rock off course before it hits the Earth; but like those imaginary space battles, these efforts would require strong forces and will happen only in the future if at all.

However, NASA also wants to pull in smaller things like tiny extraterrestrial particles that can carry valuable information. Obtaining such samples is a rapidly growing part of space exploration. NASA's Stardust space probe, launched in 1999, gathered microscopic dust particles from a comet called Wild 2, and returned them to Earth for analysis in 2006. In 2003, JAXA launched its Hayabusa space craft, which retrieved bits of a small asteroid and brought them back to Earth in 2010. And in November 2011, NASA launched its Mars Science Lab with the Curiosity rover, which will scoop up samples of Martian soil and analyze them for signs of life processes.

Using light to attract and manipulate small samples in space or from planets and other bodies would complement and extend these kinds of missions, each of which takes considerable effort and expense. Light-based sample harvesting could maybe be done at a lower cost and could also allow continual monitoring, for instance, of a planetary atmosphere. According to recent research and some older results, there is reason to believe this is not just wishful thinking.

In two separate theoretical papers published in 2011, a group at the University of Central Florida, and another from the Technical University of Denmark and the National University of Singapore, showed how to make an object move backward along a light beam. The basic idea is deceptively simple; because the incoming photons carry momentum, if you can get some of them to bounce or scatter off the object in the same direction they are traveling—that is, in the area where the object would ordinarily cast a shadow—then to balance out the momentum going forward, the object has to move backward toward the light source. This would require a laser beam with a carefully designed pattern of varying brightness across its diameter, which is not easy to create, and so these ideas have yet to be experimentally tested. But NASA has enough faith in the approach that it started to fund research in optical tractor beams.

One other potential optical method goes back almost 50 years. In 1964, Victor Veselago, of Moscow's Lebedev Physics Institute, theorized about an optical medium with a negative refractive index. In ordinary media like water or glass, the refractive index is a positive number that determines the speed of light in the medium and also how much a light ray bends or refracts when it enters another medium. Refraction is the reason that a stick partly inserted in water seems to break and bend upward at the water's surface. But in a medium with a negative refractive index, the stick would display "backward" refraction and seem to bend down.

Negative refractive indices as Veselago envisioned them have been realized in carefully designed, artificially constructed media called metamaterials. In an especially intriguing breakthrough, these ideas also led to the creation of the world's first true invisibility cloak, made in 2006 by David Smith and his group at Duke University.

Veselago predicted another "backward" result that amounts to a tractor effect, which is that a mirror embedded in a negative index material would be pulled rather than pushed by a light source. In 2009, Henri Lezec, of the U. S. National Institute of Standards and Technology, described how he and Kenneth Chau tested this theory. They fabricated a metamaterial with a negative refractive index in the form of a tiny nanoscale movable lever, and found that it was indeed pulled toward light from a laser. This would seem to be the first observation of light acting as a tractor beam according to Veselago's theory, but the results do not fit the predicted behavior in detail. The experiment or the theory may be anomalous, and the experimental result remains under study.

If metamaterials display a true tractor effect that would be fascinating, but it would not be exactly what NASA needs to gather up space debris or interesting space dust, nor would it be useful in future space battles. To reel in a particular enemy ship, some unlucky crewmember would first have to don a spacesuit, venture out, and paint the target with negative refraction paint. If that is the case, we might just as well go back to grappling hooks, ropes, and spacesuit-wearing boarding parties armed with cutlasses.

Chapter 3

Culture

Introduction

"Culture" is typically taken to mean art, theater, and music as well as pop culture, but it's more than that. It's the sum total of a way of life and modes of thinking, of the shared customs and beliefs within a human society or group. In this broader view, the practice of science is as much a part of culture and society as any other human activity. This chapter explores the science in our culture through its effects on art, the media, and daily life, and examines the subculture of science and scientists with its own rules and beliefs.

Science and scientists meet culture

"In Salmon Do Did Mobile Bond. . ." (1998) relates how I once tried to make a computer interpret an innately human form of expression by recognizing my words as I read poetry. The results highlight the gap between people and machines, which voice recognition has yet to bridge. "Laughing by Numbers" (1991) shows how quantitatively-trained scientists can find fun in the mathematics that most people find dry and intimidating.

"Real Physicists Don't Wear Ties" (1991) explores why scientists would rather not be caught dead in typical business dress. "Spelling

Real Scientists Don't Wear Ties: When Science Meets Culture
Sidney Perkowitz
Copyright © 2020 Sidney Perkowitz
ISBN 978-981-4800-68-6 (Hardcover), 978-0-429-35145-7 (eBook)
www.jennystanford.com

It Right in Karachi" (1993) explains how and why scientists' names get attached to their results, from single researchers to huge research teams (it took over a thousand physicists to discover gravitational waves; see "These Georgia Tech Physicists Helped Prove Einstein Right.") "Brother, Can You Spare a Cyclotron" (1997) from *MIT Technology Review* shows how the Great Depression enhanced the relation between science and society in the United States.

Cooking with science

One way that science enters daily life is through cooking, beginning with the invention in 1800 of the kitchen range by the physicist Count Rumford, a pioneer in thermodynamics. The involvement continued with the invention of the microwave oven in 1947, based on the development of radar in World War II. "Food for (Future) Thought or *Star Trek: The Menu*" (2011) presents further high-tech 21st-century approaches to preparing food. "The Future of Meat" (2014) explains how genetic science may enable us to grow meat in a lab rather than harvesting it from animals, with potential benefits for global sustainability.

Science and art

Are science and art really in opposition? "Art Upsets, Science Reassures" (1996) points out how these essential human activities are different but also similar. "Hubs, Struts, and Aesthetics" (1996) explores the relation between aesthetic form and useful function in industrial design. "Inspirational Realism; Chesley Bonestell and Astronomical Art" (2012) shows how stunningly realistic images of planets and spacecraft made by the illustrator Chesley Bonestell helped launch the space age in the 1950s and 1960s. "Art, Physics, and Revolution" (2014) portrays the work of the Mexican muralist David Alfaro Siqueiros and of the American expressionist Jackson Pollock in terms of the physics of fluids, and "Mr. Turner, Artist, Meets Mrs. Somerville, Scientist" (2016) shows the interaction between a great 19th-century artist and a pioneering 19th-century female scientist.

Science, literature, and the media

As an example of technology and literature, "Connecting with E. M. Forster" (1996) returns to E. M. Forster's views on technology in light of his famous literary dictum "Only connect!" Today the internet connects people in their billions, but we are only beginning to understand its deeper effects on human interactions. Forster's distrust of the technology of connectivity was expressed in *The Machine Stops*. Yet as a closeted gay man in 20th-century England, he could well have welcomed the opportunity the internet offers to safely join a supportive community, one benefit that this technology can bring.

The biggest popular impact of science is through science fiction. "Science Fiction Covers the Universe and Also Our Own Little World" (2009) shows how a futuristic film like Alex Rivera's *Sleep Dealer* (2008) can address real global issues such as immigration. "How Realistic Are Movies Set in Space?" (2014) and "Hollywood Science: Good for Hollywood, Bad for Science?" (2013) review portrayals of science and scientists in science-fiction films and how they affect citizens, science teaching, and young scientists-to-be. "Turing and Hawking, Typical Nerds?" (2015) evaluates on-screen biopics about two eminent scientists of our time, Alan Turing and Stephen Hawking (the piece was written before Hawking's death in 2018).

"Boldly Going for 50 Years" (2016) is a history of the *Star Trek* television and film series and its social impact written on its 50th anniversary. "Abstract Theory Has Real Consequences, in the Past and Today" (2017) moves away from Hollywood-scale productions to consider three smaller films created by independent filmmakers who present real science rather than science fiction.

In Salmon Do Did Mobile Bond. . .

Robby, the robot did it. So did HAL, the murderous computer from *2001: A Space Odyssey*, and C3PO, the golden fool from *Star Wars*. Even C3PO's vacuum cleaner-like sidekick R2D2 managed it, as did Gort, the towering automaton from *The Day the Earth Stood Still*— though they both hid their lights under bushels. The "it" in all this is the ability to understand language, a sophisticated brain function that, some cognitive scientists say, lifts humanity above lesser creatures.

If you hadn't noticed that these robots were so sophisticated, it could be because they were all given the strangest of vocal mannerisms. Robby, who appeared first in *Forbidden Planet*, spoke in very artificial tones, HAL sounded constantly surprised and C3PO could not stop its stream of hysterical nonsense. R2D2 communicated in whistles and burps while Gort failed to utter a single word.

Which is all rather strange in such advanced machines: each of them was at some time entrusted with caring for human life (in Gort's case, only the voice command "Gort! Klaatu barada nikto!" stopped it destroying all life on Earth). Let's face it, with the right interface, it seems likely that all of them would have passed the ultimate test of computer smarts—the Turing test. That is, they would all have given human-like answers to any questions a human might ask.

Things are rather different outside Hollywood, where computerized machines are still struggling to understand even the simplest verbal communication. And so, the instant IBM promised it was possible to "compute your way through the day with just your voice and VoiceType," its speech-recognition software, I bought the program, primed for intelligent conversation with my computer.

Intelligent, that is, after I had been properly trained. Even human linguistic ability can be stymied by our wild variety of accents and rhythms. Rather than accommodate this diversity, VoiceType tries to regularize human delivery. It taught me to speak "isolated word speech" into its microphone, saying each...word...with...a...definite... and...perceptible...pause afterwards. By the time I graduated from speech-recognition boot camp, I was droning out words in approved robotic style.

Talking like Robby the Robot was just fine when I used VoiceType for brief commands such as "Go to Media Player," which brought up that function on the computer screen. But a measured delivery was harder to sustain while dictating longer passages to create a text document. Still, as I repeated that famous alphabetical incantation, "The quick brown fox jumped over the lazy sleeping dog," the words appeared on the screen just as spoken.

All well and good. Then I began wondering whether the computer could fathom speech characteristic of humans rather than machines—the allusive, rhythmic language of poetry. Not wanting to blow the computer's mind, I started it at the low or doggerel end. First, I recited a childhood favorite: "Roses are red, violets are blue, with a face like yours, I'd live in the zoo." VoiceType handled it well, although changing "violets" to "violence" was a bit of a slip.

But then hubris set in. Like Frankenstein, I violated the natural order: I deliberately introduced poems both romantic and mystical, the very antithesis of machine style, into my computer's silicon brain. I really felt the difference as I recited works by William Blake and Samuel Taylor Coleridge, for it was a struggle to keep from breaking out of the required robotic delivery and into the powerful rhythms the poets had intended.

The visionary poet and artist William Blake was no great fan of science. In 1795, he made a print showing a naked Isaac Newton imposing a distasteful rationality on the world. But science has now got its revenge—in what VoiceType did to Blake's *The Tyger*. The original reads:

> Tyger Tyger, burning bright,
> In the forests of the night;
> What immortal hand or eye,
> Could frame thy fearful symmetry?

IBM's version—care of my voice—runs somewhat differently:

> Tiger, tiger, turning right
> In the virus of the night
> What immortal and worldwide
> Could flame thy fearful salmon tried.

Giving credit where it's due, I must say that "virus of the night" could be a dark metaphor for disease, more poetic than what most would-be poets declaim at the local coffee shop. But what on earth is that "turning right" business? Could it be a conservative political message? And try as I might, I can't connect a nervous fish or "fearful salmon" with the unutterable power of God's creation. So, despite that one viral image, I awarded the program a C-minus in comprehension.

Still, I couldn't give up just yet. So I tried another mystical work. Coleridge wrote *Kubla Khan* while immersed in the vestiges of an opium dream. And the poem does suggest an altered state of consciousness. It begins like this:

> In Xanadu did Kubla Khan
> A stately pleasure-dome decree:
> Where Alph, the sacred river, ran
> Through caverns measureless to man
> Down to a sunless sea.

That wondrous name, "Xanadu", totally baffled VoiceType. In different readings, the computer turned it into "seventy-two" or "salmon do," mysteriously rediscovering the fish theme from *The Tyger*.

Nor could VoiceType get a handle on the name "Kubla Khan," rendering it as "mobile bond" or "global kind." So the first line was ruined, and that evocative river Alph also fared badly, becoming "alcohol," "health," or "Alvin." A sacred river of health or alcohol makes a kind of demented sense, but on my list of exotic names, "Alvin" scores pretty low. VoiceType's efforts here earned only a D-minus.

All this reminds me of a science-fiction tale about a human explorer who crashes on a hostile planet. He finds a shelter with food, air, even entertainment, but it was built by and for an alien race, and is harmful to him. Facing certain death, the explorer falls unconscious, only to be roused by the delicious smell of cooking, while he breathes fresh air, and hears lovely music. Has the shelter changed to accommodate him? No. When he looks down at his body he sees the automated refuge has transmuted him into the alien form it was meant to serve.

The Turing test is, after all, a two-way street. As I represented humankind by reciting poetry, the computer represented machinekind by making me talk like a robot. So while I enhanced its humanity, it diminished mine. And heaven help us, I suspect that long before my PC types "Tyger Tyger, burning bright," I'll have come to understand just what it means when it says "In seventy-two did global kind a steeply pleasure-dome decree."

Laughing by Numbers

Have you heard the one about the group of prison inmates who puzzled a visitor to their grim penitentiary? "Thirty-nine," shouted one, and the rest rolled around in laughter "Fourteen," said another, and more guffaws ensued.

A guard explained that these bored long-termers had simply assigned numbers to their favorite jokes, as an easy way to entertain each other and pass the time. But the now-enlightened visitor noticed that one prisoner, although he shouted "forty-three" and "seven" with gusto, never got a laugh. "Oh," said the guard, "that's old Gus. He just doesn't know how to tell a joke."

I suspect those prisoners were ex-scientists (possibly sent up the river on charges of insufficient publication, compounded by felonious overhead mismanagement) since only people with intense quantitative training could get joy or any other emotional frisson out of numbers. Oh, there are a few exceptions—our ages and waist sizes probably carry emotional baggage for most of us, scientist or not. But I'm thinking of numbers with strong associations distant from such direct connections. If these don't exactly create a mood, they at least generate a striking non-numerical mental image, or unexpectedly mine a rich lode of memories or knowledge.

I first really noticed this feature of my own internal landscape one day, when I idly glanced at an auto license plate showing the sequence 488. Instantly, the picture of a sizzling blue light ray snapped into mind, like the very best special effect from *Star Wars*. From where did this spring? It's simple. I use lasers in my research, and know that 488 is the wavelength in nanometers of the bright blue line emitted by one of the more powerful sources, an argon ion laser.

Once I noticed this, I realized how certain numbers—on license plates, credit cards, in my work—bring strong imagery or ideas along with them. The host of laser numbers is especially vivid; 514, also from an argon ion laser, is a beautiful emerald green that the Wizard of Oz would love; 613, from a helium neon laser, a torchy red-orange. But 337 carries no image. More exactly, it's invisible; expressed in micrometers as the main line from a hydrogen cyanide laser, it lies far, far out in the infrared, well past where the human eye responds.

Other numbers lack the sheer visual presence of the colorful laser line set, and don't even carry physical units like nanometers, but impress through their compelling logic. Anyone who deals much with computers learns the powers of 2, even the bigger ones, such as 1024, 2 to the 10th power.

These potent numbers are also so innocently charming in the ultimately simple binary code they represent, that like a favored child, the apple of my eye who always draws attention, I always notice when they pop up in ordinary life. Because they're easy to remember, they're good codes for, say, a bicycle lock—at least if other fanciers of the binary are not in sight.

Then there are the honored loons of numbers which have been around long enough to grow intricate and layered histories. One group is the irrationals: π of course; e, the base of natural logarithms; and the length-to-width ratio $(1 + \sqrt{5})/2$, the aesthetically pleasing proportion of the golden rectangle. The stories of their discovery, the puzzlement over their unending digits, the searches to find good approximations or to add yet more thousands of digits, create a kind of awe. These numbers and others, such as various infinite enumerations, carry a vaguely mystical tinge of elusive higher truth. Perhaps akin to Platonic ideal forms, they may be as close as some scientists ever get to the spiritual.

Generating amazement rather than awe are what might be called the "isn't it a small universe" numbers of coincidence. To take these seriously is to risk the madness of numerology; still it rings in my mind that 13.6 is both mercury's density (in grams per milliliter, at 0 °C), and hydrogen's ionization energy (in electron volts); and that the Earth's mass in kilograms is very nearly ten times Avogadro's number. These also produce the unscientific but seductive thought: "Is there really—somehow—a connection?"

This is all very well, but I'm begging the question: do people, even scientists, ever actually laugh at numbers? Unlike the prisoners in the penitentiary, the answer is probably "no" for single integers or decimal numbers, no matter how deftly delivered by a raconteur of numerals; but "yes," for numerical comparisons conveying the basic comic elements of exaggeration and surprise. Absurdly magnified overstatement, restrained understatement and catastrophe, ridiculously heightened contrast, are all funny.

I would laugh if someone said, "The distance to Alpha Centauri is 4.3 light years, 2 inches," and I have snickered at Arthur Stanley Eddington's statement, "I believe there are 15, 747, 724, 136, 275, 002, 577, 605, 653, 961, 181, 555, 468, 044, 717, 914, 527, 116, 709, 366, 231, 425, 076, 185, 631, 296 protons in the Universe and the same number of electrons." The mind-bending disparity between light years and inches is what makes the first statement laughable; Eddington's sublime chutzpah in giving an enormous, untestable theoretical result down to the very last proton in the Universe is the tickler in the second. (If an actual count showed Eddington to be off by just a couple of protons, would he still get full credit? I'd certainly vote to give it, on the grounds of "close enough.")

Nevertheless, at my next physics conference, I'll try for a laugh like Gus at the penitentiary; by shouting out the right number in the right way. Logical and scientific, I'll begin with positive integers, move on to zero and negative integers, then rational numbers, irrational, complex, on to the highest reaches of numberdom. If none of these work, I'll revert to the old gags, say number 73: "An American physicist, a French biochemist, and a British chemist are sitting on the stage in Stockholm, waiting for their Nobel Prizes, when suddenly the American says . . ." But you know the rest.

Real Physicists Don't Wear Ties

I peer into the mirror as I knot my tie, a neat silk number. For a moment I enjoy its vivid red and black pattern, and then comes a flash of panic; am I ruined as a scientist, because it has become natural to wear coat and tie? Against my instincts as a paid-up member of the science tribe, this is my regular uniform on sabbatical leave in Washington, D. C. with its rigid dress code. For the first time. I directly confront an issue long simmering in my head: why do scientists dress differently from business people and from other professionals? What does this say about us to ourselves and to non-scientists? And in times when the "funding wars" need our best efforts, when scientists are seen as techno-nerds with pen-filled pockets, what should we do about it?

Clothing is a major part of how people define and present themselves. Alison Lurie puts it crisply in her book *The Language of Clothes*: "Long before I am near enough to talk to you. . .you announce your sex, age, and class to me through what you are wearing—and very possibly give me important information (or misinformation) as to your occupation, origin, personality, opinions, tastes, sexual desires, and current mood. . .By the time we meet and converse we have already spoken to each other in an older and more universal tongue."

Scientists also reveal themselves by their dress, which is light years away from Washington's formal style. Outside a military base, I have never seen such uniformity as prevails here among the capital's well-dressed male bureaucrats and lawyers. Sedate suits, inoffensive ties, and wing-tip shoes are the rule. A variation consists of a navy-blue blazer, grey slacks, a diagonally striped tie, and loafers; hardly American frontier individuality. Men's hair does not fall below the back collar line; beards are rare. Women's clothing shows slightly more range, from simple or elegant dresses to classical suits. But neither men nor women are guilty of a high-fashion look. Individual flair is overwhelmed by a universal style of well-groomed business-like seriousness. This is a town where *The Washington Post* ran a story about the shock waves that resulted when three members of Congress voted in shorts and sneakers.

On the other hand, when scientists gather, wearing sneakers is no big deal. There is no universal style. At a physics conference, men's clothing ranges from jeans—some having visited actual campfires—to formal trousers, with shirt and sports jackets, but usually with no tie. Few suits appear. Footwear is worn for comfort, with alternate use on the hiking trail a distinct possibility. Hair often extends far below the collar or out into space. The prevalence of beards—some short and neat, some of biblical grandeur—is astonishing. Two-fifths of the male physicists I see have facial hair.

The few female physicists at a conference take more care. Women are less likely than men to believe that a green plaid looks great with yellow stripes. One woman says that at a certain point in her career, she decided that good clothes were important to her and began to dress accordingly, but she still feels that too much style can threaten fellow scientists.

Dress also varies among different scientific disciplines. Physicists are especially nonchalant; one woman physicist who says she dresses for "comfort and decency" adds that "real physicists don't wear ties." Neither do they wear white lab coats, although biomedical researchers occasionally do. Outdoor types—oceanographers, geologists—bring their gear indoors, like a handsome sheepskin jacket I saw at one conference. Industrial researchers are dressier than academics. Nationality is important as well. A French physicist visiting my lab radiated casual elegance. He draped a blazer, cape-like, over his shoulders (only men born into a Romance language have the right DNA for this), with an open-collar striped shirt so nicely cut that a tie would have been a crude intrusion.

Most scientists would agree that we could never be mistaken for a group of bankers, lawyers, or yachting aristocracy. The anthropologist Sharon Traweek, who studied the culture of high energy physics at the Stanford Linear Accelerator Center, concluded in her book *Beamtimes and Lifetimes* that the physicists are dressed most casually of all staff types

> in shirts with rolled sleeves and jeans or nondescript slacks. They disdain any clothing that would distinguish them from each other. The style to which they conform, furthermore, maintains a carefully calibrated distance from fashion, quality, or fit. . .I can think of four exceptions among over one hundred physicists. [One] whose clothes are neatly pressed, well made, and color-coordinated, is said to be really okay: he only dresses "that way" because his wife buys his clothes.

Do all scientists choose to dress without style? *The Washington Post,* reporting a conference celebrating the new carbon buckyballs, pointedly noted that the attending chemists wore cheap suits that coordinated perfectly with their main fashion accessory—the black plastic digital watch.

Perhaps scientists are too poor to dress well. Scientists do not earn the highest of incomes, but plenty of Washington professionals earn even less. While few wear $1000 suits or dresses, they do dress within their budgets in businesslike style as part of the job, as I wear jacket and tie to represent my employer properly. For scientists holed up in their laboratories, "must dress well" is not in the job description. If your colleagues wear "cords. . .tie optional," or if your afternoon plans include mucking out a vacuum pump, there is no great imperative to invest in a handsome Italian suit.

Even so, there are deeper issues behind our choice. It is easy, and certainly soothing to the ego, for scientists to conclude that brilliant people searching for eternal truths need not bother with fashion's frivolities. It is not our dress, we believe, but our dedicated brainpower that leaves our mark on the Universe. Did Einstein or his admirers care in the least that the body housing this great mind wore a baggy sweater, sockless shoes, and ebullient hair?

Not long ago I saw Einstein's unruly hair used in an advertisement as a clear visual signifier for the concept "genius." The image of the almost alarmingly incisive intelligence looking outward to nature's secrets, not inward to mere details like correct clothing and haircut, is part of our mythology.

Still, there is a lurking sense that our reaction to suits—or at least expensive suits, since chemists are said to favor cheap ones—goes beyond the simple feeling that dress is unimportant. Somewhere in the scientific demonography there is a suit-wearing devil. Suits are the managerial uniform of the administrator, and scientists often give administrators special treatment that is laced with a certain disdain toward what they see as a nearly useless profession. And maybe feelings against suits well up from deeper, more hostile roots, illustrated by a planetary astronomer I know who says he's donning "battle gear" when he dresses up for a presentation to his suit-wearing bosses. It is no secret that suits can generate resentment. In his book *New Dress for Success,* John Molloy describes how male blue-collar workers "considered the suit, shirt, and tie to be a uniform that announced that the executives were superior beings," and that the workers were inferior to "the suits."

Do scientists also resent "the suits?" With our nominally high standing, you would expect us to be at ease with the suit-wearing establishment. But it is not that simple. We consider ourselves creative, intellectual professionals, yet lie outside the mold of the traditional literary intellectual. We often work with our hands, and earn less than physicians. We may be uncomfortably stuck somewhere between mechanic and philosopher, with more blue-collar in the mix than we like to believe. Or is it that, under all our smug certainty about our importance, we wonder how much we really count? The power to fund our work, to decide whether and how to use it, does not lie with us. We uncomfortably suspect that a comment attributed to Winston Churchill, "Scientists should be on tap, not on top," is gospel for "the suits" of government.

Maybe our rejection of suits carries a tinge of rebellion, the urge to prise up cobblestones and build barricades. Molloy claims that the rebellious are easily spotted through their appearance. Women with particular hairstyles, men wearing bright offbeat colors, "orange shirts, red jackets, green and yellow and purple ties," are anti-establishment, he says. If color is the rebel's indicator, scientists fail this litmus test; a science conference is a study in shades of drab.

No, any tiny mustard seeds of rebellion are expressed through hair, especially the beards popular with male scientists. According to Lurie's, book *The Language of Clothes,* a beard with long hair marks a Bohemian style. With medium length hair this modulates to "interesting, but not antisocial or disruptive, originality." Some male researchers sport these looks, but my own beard and medium-to-long hair do not say "scientist" to the world. It usually pegs me as writer, artist, or English professor. It is the beard with short hair, says Lurie, that suggests "seriousness, specialized knowledge, and often a slight inflexibility of views. The combination is often seen on research scientists, doctors, and engineers. . ." If Lurie is right, some of our beards show that we are out of society's main channel: others draw on a powerful image she calls the "old stereotype of the bearded alchemist or magician—wise but very possibly dangerous."

Beards are not the only important dress accessory. The other is the cheap pen. Usually several of these are clipped to a male scientist's breast shirt pocket in appallingly plain view. My own limited survey shows that the bulging multi-pen pocket—the magic number seems to be three—truly exists among actual scientists, not just cartoon ones.

There is no puzzle about the need for a pen, a powerful implement to understand nature. Ed Regis's book, *Who Got Einstein's Office?* describes how someone once asked Einstein where he kept his laboratory. Einstein smiled and took a fountain pen out of his breast pocket. "Here," he said. No, the issue is not "Why the pen?," but "Why so many?" It turns out there is a precedent in recent Chinese history. As Molloy relates: "When Chairman Mao put everyone in Mao suits, his object was to wipe away all signs of class [such as] clothing. He did not completely accomplish his goal, even at the height of the Cultural Revolution: Chinese army officers wore the same outfits as enlisted men but carried pens in their pockets, and if you knew how to read the code you could tell their rank by the number of pens. . ."

To an open mind, the match between the Chinese army and the science brigade is positively uncanny. According to Traweek, scientists' dress reflects their belief that they are part of an egalitarian meritocracy: "The assumption is that everyone has a fair start. This is underscored by the rigorously informal dress code." But no group can operate without separating the leaders from the led—hence, highly visible pocket pens.

This gives a male scientist two mutually exclusive ways to pull rank: he can wear a suit, covering up his pen collection; or he can throw away his suit jacket to unleash the power of his pens. Female scientists suffer in their choices for insignia of rank. They can wear suits, but not power pockets. They have one other option: female colleagues tell me that choice of jewelry communicates nuances of position and success to other women.

Scientists' dress matters more in technically based businesses. For example, the image of massive, unswerving IBM versus limber, innovative Apple Computer correlates with the difference between a buttoned-down style and the jeans–shirt look. A conscious decision on style went into forming another big technical company, Texas Instruments (TI). Pat Haggerty, an ex-naval officer and founder of TI, thought that a uniform look would strangle creativity and repel the kind of people he wanted. He and other founders wore open collars. Over at IBM, ex-salesman Tom Watson ordained conservative dress to show seriousness and reliability.

Scientists need both the IBM and the Apple–TI looks. It is easy to tell when each is right. Among ourselves, we should never slavishly follow anyone's rules; it we feel that a tie or scarf cramps creativity—perhaps by reducing the brain's blood supply—we should dress as we like. But at the interfaces between us and the rest of the world, it hurts if we project a head-in-the-cosmos image. To look IBM-serious is a good goal when teaching, explaining, or selling to people who do not know us all that well.

One way to accomplish this is with hired guns. Lobbyists know how to wear the right clothes when they sell science or anything else. But scientists themselves could dress to persuade using an approach I found at a college in the Virgin Islands, where people do not need winter clothes. When a science faculty member had to brave northern cold to raise funds, he wore the departmental overcoat of intermediate size which more or less fitted everyone. Our professional societies could easily provide male and female campaign suits for the few times when they are needed.

What is really wanted, however, is a radical, distinctive form of dress that expresses the essence of our calling. Molloy comments that physicians are notoriously poor dressers, yet their working uniforms radiate power. All-white hospital wear is "a high-authority outfit"—the doctor's white jacket "makes whatever he or she says and does appear all but infallible." Leonard Laster writes that these costumes work because we once believed that the medicine man could "change the course of sickness and slow the approach of death. Costumes and masks strengthened the belief. . .the white coat [symbolizes] the myth about the magic of medicine."

A little awe, based on our own mythology, might work wonders in the halls of power or in attracting neophytes to our profession. My idea is that we adopt an impressive science uniform to identify us as members of a dedicated calling with ancient roots. The right design could serve both men and women and provide a set of egalitarian power insignia.

Here is what I envisage; for comfort, a bottom layer consisting of white sneakers and sweat socks, denim jeans or skirt, and white T-shirt. Over this, my radical breakthrough, bestowing dignity, mystery, and an aura of unknown powers—the science robe. Think of Merlin the Magician, or of the Sorcerer's Apprentice sequence in Walt Disney's animated film *Fantasia*, where Mickey Mouse and his boss appear in full magical wear. Their robes, updated to represent modern science, rather than ancient wizardry, would radiate all the power and distinction we could want.

How would we update them? My memory is that Merlin's model was covered with stars and other astrological signs. It would be easy to design a dark blue robe (70 per cent wool and 30 per cent polyester for quality and wrinkle resistance, in lengths and weights suitable for summer or winter wear), peppered with golden symbols representing the best of modern science: a DNA molecule, a semiconductor superlattice, a quark, a fractal image, a black hole, and so on.

For individuality, a discreet shoulder patch would give the wearer's scientific specialty. To continue the tradition of pocket power, the robe would carry a mock breast pocket complete with embroidered pens. Armies and airlines count seniority with stripes; a golden circle for each five years' service stitched around each cuff of an officer's uniform coat. Similarly, each pen would represent five years of scientific experience. Borrowing another old military tradition, missions accomplished would also receive their due: a small image of a research journal for every 10 papers published; a dollar sign to mark every half-million dollars of accumulated grant support; and a stylized human figure to indicate a student guided to a doctorate, with added star if he or she actually got a job.

Just imagine the stir as an eminent scientist, called to testify before Congress on a matter of national urgency, sweeps into the hearing room in full regalia, robe billowing, his or her career proudly outlined on its chest. People would say, in hushed tones. "Here comes a Scientist" with a capital S; the sound bite on network evening news would be unforgettable; and thousands of young people would discover a compelling urge to enter this dramatic profession.

Can we possibly pass up this opportunity to influence, to recruit, and to convey our sense of wonder about the world? I urge that we trade in any business wear we own—I'll sacrifice my stylish red and black tie—and our dingiest jeans for these benefits, and the chance to peer in the mirror without self-doubt.

Spelling It Right in Karachi

Like many another scientist, I have long dreamed of leaving my mark on my calling. Short of a Nobel Prize, the next best bet would be to firmly tattoo my name onto a startling discovery or original idea. The Perkowitz shift, say, a new physical effect; the Perkowitz equations, an incisive summary of a complex set of phenomena; or even the Perkowitz conjecture, a brilliant speculation that creatively galvanized a whole field—all seem worthy goals, enriching science while pleasing me. This would only follow precedent, which has long linked scientists' names to their breakthrough science. Our textbooks preserve Avogadro's number, Mendel's genetics, Newton's laws of motion, and so on. Each of these researchers presumably now sleeps the sleep of the deeply fulfilled.

I've had my name polished and ready to go for some time, awaiting only the world-class notion or discovery, which—it is easy to forget—does need to come first. Unless the idea is important, it is not worth the lengthy process of adding the discoverer's name. In the past, after the theory was proposed or the new effect reported, the scientific community checked out its validity. Only when the insight proved valuable did it became an honor to attach the discoverer's name to it. This verification took time. More time was needed to determine who actually deserved the credit, even in days past when more scientists worked alone. This may explain why we see fewer scientists' names in these days of fast-breaking results and large research groups. Instead, the hot ideas in science—chaos, the science of complexity, the Grand Unified Theory, black holes—seem to be named on a principle other than recognition of individual achievement. At one extreme, it may be akin to the primitive tribesman's faith that the name is the power. Even in our modern world, advertising people know that calling a sports car "the QZ2700" rather than just "fast automobile" gives it a unique and rakish air.

More charitably, some current titles show the brilliant compressive force of poetry, like Stephen Hawking's "Big Crunch" for the end of the Universe. Others carry effective, accurate imagery, such as "greenhouse effect" for global warming. Sadly, some display a propensity for lame science jokes (does the Universe really need a fundamental physical property called "charm?"). Another possibility is that this new way of naming reflects the importance of research teams. A recent paper about elementary particles lists 104 authors. Another in the life sciences is burdened with 193 names. However important these results may be, neither could carry such a mob.

The tribesman and the copywriter are not wrong. Titles are important, and unexpected things happen from our process of naming. One example comes to mind from Nathaniel Hawthorne's classic novel *The Scarlet Letter*, which most Americans of my generation read at school. In the story, set in America's early days, the adulterous Hester Prynne is shamed for her sin. She is forced to display on the bosom of her dress a large, damning red letter "A." A quaint old tale, you might say, with no earthly connection to modern science. But the pull of scientific poetry is strong, for not long ago at a literary conference, I heard poor Hester's symbol called a "strange attractor." And the attention given to the science of chaos, as immortalized in James Gleick's book, reminds us of another aspect of naming: the complexities of burdening common words with additional specific scientific meanings.

So our scientific names have immense power. We should not abandon the vital motivation of personalized achievement, but need to add modern public relations. The world of rock music can show us how. Consider, for instance, James Clerk Maxwell, the 19th-century mathematical physicist who derived the four equations of electromagnetism, which predict the speed of light. Suppose he were a young scientific newcomer today. To start building fame, we might distribute his early calculation that showed the rings of Saturn could not be solid, the equivalent of a good single cut that caught disc jockey attention. Other similar rising successes would follow, enlivened with carefully timed snippets about his romantic entanglements. Creative styling of his beard could suggest both scientist and visionary. He would not actually have to play guitar, but the hint that he did so in time off from theorizing wouldn't hurt, like President Clinton's involvement with the saxophone. Soon, in rising anticipation, scientists and the whole world would eagerly await Maxwell's Equations. When they finally appeared, they would instantly go platinum, like any Michael Jackson album. At this point in his career, Maxwell could drop the "James Clerk," having reached the Madonna-like stage of hyperfame where one name suffices, and without ever having posed naked.

A very different use of the primitive power of names has been unaccountably overlooked. If we pick the correct title before the discovery is made, we can avoid dead ends and help the researchers get it right. It is, for example, an exciting goal to derive the Theory of Everything, which explains the four forces of nature. But scientists trying to meet this tall order are under enormous pressure. What if they get it only nearly right, leaving large unexplained chunks of the cosmos as intellectual orphans? Imagine how life would be simplified if we boldly announce instead a search for TONE, the Theory of Nearly Everything. TONE would be required to explain all physical phenomena, and all biological behavior up through the cellular level; but would be permitted, even encouraged, to leave as blessed mysteries the stock market, human love, and the reason the telephone always rings when one is in the shower.

There is one final overriding need in naming science: simplicity. Humphrey Bogart, who ought to know, once said "You're not a star until they can spell your name in Karachi." But we complicate the spelling by insisting on the likes of the "Schrodinger equation." Our best policy may be to limit our famous names to four or five letters, giving James Watt, Ernst Mach, Francis Crick and Niels Bohr central places in our pantheon. This would, unfortunately, also eliminate my own nine-letter name, but I can always change it to "Park" when my big moment comes.

Brother, Can You Spare a Cyclotron?

I missed the defining event of my parents' generation, the Great Depression. But I heard plenty about it, through tales about the jobless selling apples on the street and the songs Woody Guthrie sang. And eventually I found that so large an event leaves other traces for later generations to unearth. Some are tangible, such as the facilities constructed under the Works Progress Administration (the WPA, later called the Work Projects Administration), which President Franklin Roosevelt founded by executive order in 1935 to employ millions of the jobless. If you have flown into Washington's National (now Reagan) Airport, driven down Manhattan's East Side Drive, or used any of thousands of other facilities, roads, and buildings, you have encountered this physical legacy.

Less concrete, but with its own kind of permanence, is the cultural legacy of the WPA, which employed artists along with the legions of blue-collar workers needed for construction jobs. Theatrical productions like Orson Welles's *Macbeth*; murals that decorated (some said blemished) the New York Public Library and other buildings; art prints and posters; American travel guides— these and more came out of the WPA Federal Art, Theatre, Music, and Writers' projects. And the WPA supported artists who would later do important work. Richard Wright began *Native Son* in time allowed for creative writing apart from his WPA assignment; Jackson Pollock started developing his abstract style while receiving a WPA paycheck.

Not so well known is the legacy left by the scientists and engineers who conducted research with WPA help. It too includes famous practitioners and classic works: research by the likes of Glenn Seaborg, who won a Nobel Prize in 1951 for discovering plutonium and other atomic elements beyond uranium; significant compilations of data such as the MIT Wavelength Tables, which became a Rosetta stone for scientific research; and experiments conducted on atom-smashing machines, the Big Science of the time. Not only do these moments in American scientific culture deserve recollection, but a look at WPA science, as well as some of its other endeavors, offers lessons for today. Although we 1997-model Americans have it far better than did the Depression generation, many scientists also operate in a climate of scarcity—layoffs and shrinking budgets—reminiscent of those earlier days.

Keeping the Science Gears Turning

Federally supported science was not utterly new in the 1930s. The public had long valued and paid for agricultural research; military technology received funding during the Civil War and World War I; and since 1901 the government had maintained the National Bureau of Standards (now the National Institute of Standards and Technology). But this support did not extend to broad-based, long-term, non-directed research, the kind professors perform at universities. State funds supported such efforts at leading state universities, but most funding for pure research came from private foundations. Corporations such as the American Telephone and Telegraph Co. funded applied research in their laboratories, figuring the work would lead to profitable new products.

When the Depression decimated these research funds, a group of leading scientists, headed by Karl T. Compton, then president of MIT, asked for a federal investment of $75 million over five years—at the time an enormous commitment—mostly for university research. The federal government rejected the request because the scientists were unwilling to specify how the money would be spent; instead, the government asked WPA to contribute to science and engineering research by paying for the assistants and support personnel needed to work with academic scientists. In fact, the WPA put nine-tenths of its science budget into salaries for relatively untrained people before the agency shut down in 1943. By then federal dollars were flowing to science as part of the war effort, most notably through the Manhattan Project.

Although WPA support for science amounted to only 3 percent of the agency's overall funding, the funds were 10 times greater than those for either the Art or Writers' projects. And a few percent of a budget of some $14 billion over the lifetime of the WPA was enough to influence a great deal of research. The WPA Index of Research Projects through mid-1939 lists 60 efforts in mathematics, physics, chemistry, and astronomy, more than 300 in biomedical science, and hundreds more in other sciences and technology. Much of this research was of publishable quality: two-thirds of the projects in physical science, for instance, were reported in journals like the *Physical Review*, which is still preeminent.

Familiar names appear on some of these articles, including those of three outstanding researchers at the University of California, Berkeley, who operated on the cutting edge of nuclear physics with the assistance of WPA financing for their staff. Glenn Seaborg bombarded atomic nuclei with subatomic particles to transform one kind of nucleus into another, work related to his Nobel Prize–winning research. Luis Alvarez, who was to win the 1968 Nobel Prize in physics for his method of detecting elementary particles, explored how an atomic nucleus captures its surrounding electrons, a process that illuminates the theory of antimatter. Ernest Lawrence won the 1939 Nobel Prize in physics for his invention in 1932 of the cyclotron, the first truly powerful atom smasher. In two WPA-supported projects, Lawrence used the device to make neutrons, which had been discovered in 1932, and tested their power to destroy tumors.

Other projects led to compilations of archival data with enormous scientific impact. The MIT Wavelength Tables did nothing less than determine the characteristic light emitted by each element in gaseous form and at high temperatures—hydrogen, oxygen, and the 100-odd others that make up the universe. For example, hydrogen emits ultraviolet radiation and mercury a red glow under those conditions. Knowing the wavelengths of the light allows scientists to unequivocally identify the matter that emitted it, no matter how far away. What we know of the birth, death, and constitution of stars comes from analyzing their light.

Similarly, the WPA Mathematics Tables Project, conducted in collaboration with the Bureau of Standards, extended the quantitative language that scientists use to describe the world. Over centuries of analysis, certain functions have proven essential to the mathematical vocabulary. For example, the recurring peaks and valleys of the sine wave describe the repetitive vibrations common in nature that form waves of water, sound, and light. Other examples include the exponential function, which describes extremely rapid physical change, and Legendre functions, named for the 18th-century French mathematician who first explored them, which describe electric fields and quantum behavior.

In 1938, when the mathematics project began, its aim was to calculate these useful functions and publish the results in tabular form. A contemporary article characterizes the project's computational facility as the largest ever established. It used some 150 electrically powered machines, which added and subtracted numbers the same way an automobile odometer works, with rotating gears whose positions represent numbers and interlock so that results can "carry" from one column to the next. Some 250 WPA-supported staff worked among the slowly churning electromechanical monsters, with their characteristic "chinga-ching" sounds. The employees, who were known as "computers," checked the results and transferred them from one machine to another, since no single machine could calculate the functions all by itself. The effort continued from 9 a. m. to midnight five days a week, year after year. By 1942, the project had published 12 volumes of tables and had also performed secret military calculations.

Today, of course, printed mathematical tables no longer enjoy brisk sales; people can instantly determine functions using calculators and electronic computers. But during the WPA era-before 1944, when Harvard University researchers built the electromechanical Mark I computer, which followed stored instructions, and before University of Pennsylvania scientists completed the first programmable electronic digital computer in 1946, the development of the tables promised a significant boon to science.

PhDs Aren't Everything

The Mathematics Table Project and other WPA research efforts remind us that although science needs people with doctorates, it also needs those trained less intensively. If we make earning a doctorate the only worthy goal of scientific education, we may not best serve the long-term interests of science and those who are drawn to it. The issue is relevant now because we may be producing too many PhDs in the sciences. In physics, for instance, we annually turn out 1,400 new doctorates for 700 positions. The WPA focus on support staff for scientists suggests a way out of this bind. People with good technical, bachelor's, and master's degrees—not just those with doctoral degrees—also play important roles in launching a Hubble Space Telescope or turning an idea into a product.

We should avoid making hasty decisions about the numbers of PhDs needed, since employment prospects can change over the years necessary to produce a scientist with a doctorate. But the more choices we can give students during this uncertain funding period, the more they—and science—can succeed. One way to achieve such flexibility is to offer a variety of degrees.

The Physics Department at Emory University, where I teach, offers both a traditional BS degree and a BA. The latter requires fewer physics courses and is intended for those who want to pursue directions other than a graduate physics program. We also offer a BS in applied physics, representing a move away from the idea that every student must study advanced quantum mechanics. This curriculum trades some standard courses in theory for others in optics, computing, and electronics, preparing students for either immediate employment or graduate work in those fields. The applied track has become our department's most popular undergraduate program.

Graduate education can also be made more flexible by offering highly specialized master's degrees that are not just traditional low-value whistle-stops along the track to a doctorate but provide substantial training in, say, growing semiconductor materials. And if graduate education were to include more practice in writing, speaking, teaching, and managing research, it would give students additional abilities to help them keep up with a changing job market.

Another lesson from the WPA era is that, although science needs proper facilities, a healthy scientific enterprise can continue even in the face of government cutbacks in funding for equipment. The WPA paid for people instead. Scientists such as Lawrence, Alvarez, and Seaborg made do with existing facilities or exercised their ingenuity to find other sources of support, such as the nonprofit Research Corporation. (Since 1912, this nonprofit foundation has applied the proceeds from an invention that reduced industrial air pollution for the "advancement of technical and scientific investigation"—varied research that has, in fact, included some of my work.) Lawrence also raised about $2 million from the Rockefeller Foundation and other nongovernment donors to begin building, in 1940, the world's biggest cyclotron, then the pinnacle of elementary particle research.

Today private funding still has its impact. For instance, the W. M. Keck Foundation has given $140 million to build an observatory housing an immense telescope on the extinct Mauna Kea volcano in Hawaii. But the costs of many kinds of equipment have outstripped the reserves of private support—and sometimes even government aid. In 1993, the Superconducting Super Collider, descended from the cyclotron, had already cost $2 billion when the federal government abandoned preliminary construction in the Texas desert rather than spend another $9 billion.

While scientists should continue seeking and receiving federal money to support needed equipment or upgrades of valuable but aging facilities, they can also try to replace dollars with ingenuity, as NASA scientists and engineers have already done. For example, they simplified the large Cassini spacecraft launched in October, 1997 to examine Saturn and its environs. One change eliminated a rotating platform that was to hold astronomical instruments. Without the platform the space vehicle must alternate between gathering data and turning its body so its antenna faces the earth, which enables the craft to send home the information. Still, the device can harvest a broad range of information. Such modifications have reduced costs of the Cassini mission by one-fifth.

And in elementary particle physics, costs will be shaved from the next huge accelerator, the Large Hadron Collider, because it will be built within an existing tunnel some 17 miles around. And because that tunnel is located at the European Laboratory for Particle Physics (known as CERN), the international agency for particle research that straddles the Swiss–French border, support should be readily available from several nations. Researchers are also beginning to examine novel and potentially much cheaper table-top-size, laser-based techniques that may someday serve to raise elementary particles to high energies.

As scientists face funding realities, they also need to confront an inevitable corollary: if science needs public dollars, it must win public acceptance. That means showing that the work is important to society. The WPA offers a lesson here as well, but through its artistic rather than scientific activities. Poring over WPA reports, I found no efforts to present science to the public, although scientific breakthroughs did attract popular attention. But the WPA made a point of bringing its artistic activities to people. The Music Project invited Aaron Copland and Virgil Thomson to conduct public concerts; the Art Project attempted to beautify the civic world. These efforts were not meant to turn most citizens into painters and composers but to show that culture is, or should be, part of our lives.

In 1997, science too is part of our lives. It has become an economic engine. Yet its practitioners often fail to impart to the public a sense of how science works and what it has accomplished; they fail to awaken the sense of wonder that occurs when we gain insights into the human mind or find planets beyond our solar system.

Even in the college classroom, where we are supposed to be reaching people, we often do a poor job. Few science courses and textbooks aim at non-majors. I teach astronomy to nonscientists and find that most available texts cannot bear to omit any facts whatsoever. The poor students, who have no plans to become professional astronomers, lose sight of beautiful ideas in thickets of detail. If science were more accessible, helping students to understand the natural world and their own civilization, it would yield better-informed citizens who might listen carefully when scientists ask for funding.

In the 1930s and '40s the massive WPA effort, including its science program, created jobs and thus helped hold together an unraveling social fabric and give people hope. Only if scientists understand the realities of the 1990s—only if they appreciate that science is deeply rooted in a society and an economy in good times and bad—and they respond appropriately, will society support them with a similar powerful conviction.

Food for (Future) Thought or
Star Trek: The Menu

Ever wonder how science-fiction heroes, busy hurtling through space or battling evil robots, find the time to eat nutritious meals so they can keep going? The classic science-fiction solution is that in the future, they'll take in a day's nourishment by popping a pill crammed with the Recommended Daily Allowances for all the food groups, along with pleasing taste and a satisfying sense of fullness. One little tablet, and your favorite hero is good for another 24 hours of action.

Although no one has yet managed to pack 2,000 flavorful calories into a pill, NASA tried something like that for its first astronauts. According to the space agency, the only nourishment those intrepid adventurers got were "bite-sized cubes, freeze-dried powders, and semi-liquids stuffed in aluminum tubes." Sounds delicious, doesn't it? Fortunately, NASA's space catering later reverted to more traditional foods so astronauts didn't have to make this particular sacrifice for their country.

But if NASA hasn't come up with tomorrow's ideal food, science fiction has brought new approaches to nutrition, though few seem like improvements. As usual, H. G. Wells was a pioneer. In *The Food of the Gods* (1904, film versions 1976 and 1989), chemists develop Herakleophorbia, the perfect nutritional formula for Hercules-like growth and strength. Alas, in true calamitous sci-fi style, it escapes the lab to produce monster rats, wasps, and worms along with a bad human outcome—all-out war between the Children of the Food, who are 40 feet tall, and regular people.

Then in 1953, in their classic science-fiction novel *The Space Merchants* set in a future hugely overcrowded Earth, Frederik Pohl and C. M. Kornbluth took nutrition in a different direction. They introduced Chicken Little, an enormous pulsating chicken breast 45 feet across that thrives and grows in a tank on a diet of pond scum. In what may be the least appealing "presentation" ever in the annals of fine dining, workers regularly hack off slabs of Chicken Little's protein to feed the hungry populace. Rounding out the meal, people drink a concoction called Coffiest, which seems to be to real coffee what truthiness is to truth.

Though a lump of Chicken Little and a mug of Coffiest are terminally unappetizing, what's served up in the film *Soylent Green* (1973) has to be the most repellent future food yet. In that story of upcoming global devastation, seemingly innocuous green wafers are produced in their trillions in automated factories to feed a world starved for natural foods. Unfortunately, they turn out to incorporate—brace yourself, there's no nice way to say this—dead people. Ick.

But even cannibalistic Ritz crackers have competition in the foods that alien creatures supposedly like, a favorite topic of science fiction–oriented foodies. One *Star Trek* web site lists items from nearly two dozen alien cuisines that appear in various installments of the series. The tasty Klingon treat *gagh* consists of live, squirming life forms called serpent worms, whereas the Ferengi favor equally yummy larva-like tube grubs, also served live.

These alien foods are loathsome but at least they're natural and unprocessed. Science fiction suggests the opposite for humans, projecting that we'll be eating ever more artificial and highly processed food in the future. This reflects the ongoing evolution of food and kitchen technology. For instance, after cooking over open flames for millennia, humanity moved away from fire with electric stoves in the 1920s and with microwave ovens in the 1970s, an outgrowth of World War II's radar technology. Science in the kitchen has even produced "molecular gastronomy," where chefs use liquid nitrogen, insulating foams, and other laboratory methods to produce exotic flavors and textures. And going yet further, we can now fundamentally change what we eat in the form of genetically engineered food.

But as the science-fiction examples show, excessively high-tech food processing can be deeply unappealing. Consider the "food printer," a form of digital kitchen technology promoted by Columbia University's Creative Machines Lab. This is like an inkjet printer, which works by squirting ink droplets onto paper in patterns under computer control. Replace the ink with various foodstuffs, currently anything that can be squeezed out of a syringe such as melted cheese or cake batter, and you're ready to print edible things. Instead of recipes like the hand-written ones passed down from grandma, computer programs let you construct and tweak your biscuits, say, just as you want—if your appetite can survive watching them being slowly excreted out first.

At least one chef, Homaro Cantu of Chicago's Moto Restaurant, who does inkjet printing of sushi, thinks that such printers would alter more than how we cook; they would change food distribution. "Imagine," he says, "a 3D printer making homemade apple pie without the need for farming the apples, fertilizing, transporting, refrigeration, packaging. . ." But this seems more science fiction than real science or real cooking, since what goes into even a computer-generated pie still has to come from somewhere. And how many of us would call a pie sprayed from a printer "homemade?"

Still, with some imagination, I can see unique possibilities for the food printer. Picture loading one of its reservoirs with chocolate chip cookie dough, and one with fresh, wriggling serpent worms. Then program the printer to extrude and bake the first, and to gently extrude the second just as is. Now you're all set to throw a party for both your human best friends and your favorite Klingons.

The Future of Meat

Though the United States has its share of underfed people, the average American eats more than most of us need. That isn't true for much of the Earth's population, however, which is why the world faces a food crisis. Stanford University biologist and ecologist Paul Ehrlich, known for *The Population Bomb* and other books, recently said that humanity will be taking the "greatest gamble" ever about its future if it ignores this problem. He is one of many who think action is needed—activists, researchers, and food producers—all driven by factors from long-term concern to corporate pragmatism. Much of it is aimed at the role of meat in the global diet.

Of the seven billion people on our planet, almost 900 million, mostly in underdeveloped areas, are hungry and undernourished. Two billion more lack important vitamins and minerals. The future will look even bleaker as the global population climbs to a projected 9.6 billion by 2050, according to the United Nations. With this nearly 40% increase, and as consumption grows in developing parts of the world, we will need 70% more food than we now produce.

We are running out of resources to grow and distribute food, however, as land, water, and fossil fuels become scarce, depleted, or costly, and as increased food productivity from the Green Revolution levels off. The latest report from the United Nations Intergovernmental Panel on Climate Change (IPCC) emphasizes that these problems are worsening as climate change reduces crop and seafood yields—and in a true catch-22, agricultural greenhouse gasses themselves promote climate change. Other factors are waste (the United States wasted 141 trillion calories in 2010, enough to feed those hungry 900 million for months), the inequitable distribution of food, and food's rising cost.

There is no single solution but we could move toward a more sustainable—that is, ecologically balanced and enduring—system, some argue, by eating less meat from animals. Various studies show that raising meat animals has a much bigger environmental impact than the alternatives, which include growing edible plants rather than crops to feed animals we then eat, or growing meat in the lab.

Bioscientist Isha Datar, a passionate champion of food sustainability through the non-profit New Harvest institute, which she directs, says that doesn't mean veggie burgers will or should completely replace beef burgers. In a TEDx talk on the subject, given in October of 2012 at an event in Toronto, she mixes the global with the personal, saying, "I love to eat meat. For me it is precious because it's the main feature. It defines my meal. But it's also precious because a lot of time and resources go into it." She explains the big environmental consequences of raising meat animals, and adds that changing how we eat meat is probably the "most significant and urgent step we need to take toward eating more sustainably, [but that] doesn't necessarily mean committing to veganism or vegetarianism."

Though vegans and vegetarians would welcome a further emphasis on plant-based foods, and though there is evidence of health benefits associated with eating more plants, most people, like Datar, love meat. Americans eat lots of it, and inhabitants of developing regions increasingly want more of it. Attempts to improve the food system must face this reality, as well as the status quo represented by the huge livestock industry, with its $85 billion in annual beef sales in the United States alone.

But how to institute these improvements globally? As Mark Eisler of the University of Bristol in the United Kingdom, lead author of a recent review of livestock sustainability, wrote in an email, "Translation of research results into policy and practice is as challenging as the research itself." One possibility is corporate change, if only to improve public perceptions. McDonald's, which sells billions of hamburgers a year at 36,000 outlets worldwide, announced in 2018 that it would start serving sustainable beef in Canada. Critics object to the company's definition of "sustainable," but Eisler sees it as "at least a start and open to public comment for a proposed redraft."

Another way to deal sustainably with people's desire for meat is to develop truly satisfying plant-based alternatives. Once pigeonholed as for vegetarians and vegans only, these foods are now also seen as environmentally desirable. Sales in the United States are growing, but remain only a tiny fraction of the beef market, given they lack meat's sensory appeal—not just in taste, but texture, since most meat comes from animal muscle, whose long strands of protein provide a distinctive chewiness people crave but do not get from plants.

Beyond Meat, a California start-up, is tackling that problem with support from investors like Microsoft's Bill Gates. The company's first product is a vegetable version of white meat chicken that uses "high moisture extrusion," as further developed by food scientist Fu-hung Hsieh at the University of Missouri.

I visited Hsieh there one afternoon, meeting him after he finished teaching class. Soft spoken and in command of the subject, he thinks people will always want the experience of meat. To be successful, plant-based substitutes must give them that. But meat texture, he explained, is harder to reproduce than flavor, and though the widely used soy product texturized vegetable protein (TVP) has high protein content, it lacks the familiar "mouth feel." To do better, Hsieh turned to extrusion cooking, a method used to make breakfast cereals, in which a food product is pushed through an opening by a large rotating screw assembly under conditions of heat and pressure. This process alters the food structurally and chemically—that is, cooks it.

Hsieh, with his co-researcher Harold Huff, optimized temperature and moisture content during extrusion to produce soy with long meat-like strands of protein, with further research showing that the extruded soy maintained its nutritional value. But all this science would not matter unless it offered better eating, so the effort included tastings by experts who found the stranded soy "cohesive" and "chewy," just what is wanted in a meat alternative.

Beyond Meat uses this method to make Chicken-Free Strips, sold at Whole Foods and elsewhere. Their chewiness and shredding behavior impressed critics like Food Network chef Alton Brown and New York Times food writer Mark Bittman. Though some tasters detected vegetable undertones, Bittman could not tell the Strips from the real thing when mixed with other ingredients, which was my experience when I tried the Strips at home.

In March 2014, I visited the hangar-like Beyond Meat facility in a commercial zone not far from the University of Missouri campus in Columbia. I found a busy scene, the offices having a functional look that signaled a fast-growing, "no frills" phase for the company, and plant manager Valun Singla's tour took me through unfinished areas with concrete floors scheduled for plant expansion. After donning smocks and hair protectors and stepping into a foot bath, as is required for food production areas, we walked out onto the production floor to see a lot of stainless steel machinery dominated by huge extruders, which take in soy and pea protein through giant funnels to produce the basic Strips according to Hsieh's method.

In another production area, the company is expanding its line with Beef-Free Crumbles, based on extruded pea protein, which looked remarkably like a mound of ground meat when a floor worker proudly showed me a pile just out of processing, indicating they tasted good too. Though a long way from faux steak, this could be a step toward a veggie burger, according to Singla, "even better than a hamburger."

If creating a burger is the gold standard, lab-grown meat has already made its mark, having given the world a beef burger that never saw a living cow. The technology uses methods developed for biomedical use to make "nutrition cultivated from animal stem cells and harvested independently of the growth and slaughter of animals," as defined by food scientist Mirko Betti at the University of Alberta. Paradoxically, this potential solution for our earthly problems began when NASA sought ways to feed astronauts in space. The agency funded researchers at Touro College in New York to explore the science-fictionish idea of synthetic food, and in 2002, they reported that they had multiplied muscle cells from goldfish (which are basically carp) in a nutrient bath to produce fish filets.

Though NASA support lapsed, other research continued. The biggest splash in lab-grown meat came in 2013, when Mark Post, a physiologist and physician at Maastricht University in the Netherlands, created enough beef for a meal, if a skimpy one. Post starting growing synthetic meat almost by accident, he told me, his research focused elsewhere until the manager of an ongoing project to develop "cultured meat" fell ill. Asked to take over, Post was "immediately much more enthusiastic than any of the other participants to really create cultured beef."

His enthusiasm paid off when his lab extracted stem cells from cow muscle at a slaughterhouse and nurtured them in a nutrient solution where they reproduced in the form of tubes. These were then stretched between anchor points, which induced them to "bulk up" into muscle fibers like those in natural meat.

Post estimates that, ideally, a single cell could yield over two hundred pounds of beef, though it took months to generate a modest amount at this initial stage. His team grew and painstakingly assembled thousands of strands of beef into three ounce patties, so-called "schmeatburgers," which were pan fried and served on buns to two experienced tasters at a widely covered media event in London. The tasters liked the texture, calling the flavor "somewhere between not bad, okay, and decent." The main criticism was dryness—ironically, pure lab protein lacks the fat that can be harmful to health but also gives natural beef its juicy appeal.

Fixing this lack, even turning it into an asset, is Post's next order of business. He told me he is "working on tissue engineering fat," as well as "toward getting fat with more omega-3 and -6 fatty acids," so that besides enhancing its appeal, his beef is healthier, with added polyunsaturated compounds that protect against heart disease.

As Isha Datar of New Harvest notes, though, "At present, cultured meat needs a lot of help at the basic research level." That includes funding the work, which does not come cheap. Post's schmeatburgers cost about $340,000, provided by Google co-founder Sergey Brin. The cost per pound would drop dramatically for large quantities but more research is needed to reach that point.

In contrast, plant-based meat alternatives are already consumer products that fit into a "more vegetables" strategy for the food crisis, endorsed by Paul Ehrlich, who would like to "get people to gradually change their dietary habits." Surprisingly, Post agrees, telling me, "A global shift towards vegetarian diets will solve many problems in a more efficient way than cultured beef." But, he adds, people will also want real meat, citing "the middle class in India and China, who for the first time in their lives can afford the status symbol that meat is." That is where cultured meat could come in, though consumer acceptance may be an issue.

Meat alternatives and cultured meat are not rivals. As Datar says, "Plant-based alternatives, and cultured meat, and other alternatives that we can't even imagine yet should all be considered part of a portfolio of solutions. There is no reason to suggest one will be better than another."

Like climate change, the food problem may not seem urgent to some, but the United Nations, other agencies, and researchers are taking it seriously. However, a full scale global will for change is yet to be seen. As Mark Eisler says, it's hard to turn research results into practice and policy.

Still, when I asked Isha Datar about the prospects for properly feeding the world by 2050, she drew on other experiences with sustainability to give an encouraging answer. "Mankind," she said, "has seen a lot of sectors become more resilient thanks to diversification. Look at energy and how we moved away from coal by developing wind, solar, and nuclear. You create resilience in the system by having many options."

Art Upsets, Science Reassures

When Georges Braque and Pablo Picasso together invented cubist art, they sometimes made works so similar that it is nearly impossible to tell which one was the artist, as in Braque's *Still Life with Banderillas* and Picasso's *Still Life with a Bottle of Rum* of 1911. They also shared a taste for pithy expression in words. Picasso made highly personal epigrams, such as his "I do not search, I find;" or used terse verbal comments to deflect requests for illumination, as in his dismissive response "Don't talk to the driver." Braque's flair was for philosophical aphorism, and expressed itself regularly in written form. His *Cahier de Georges Braque 1917-1947* presents nearly a hundred of these brief provocative statements, each with its black and white drawing, taken from notebooks that span his career [1].

Several of Braque's *mots* deal with science and its relation to art. I like best this one: "Art is meant to upset, science reassures," because it is sufficiently opaque to be richly suggestive. The wording suggests that "reassurance" is not an overly desirable commodity, a reading strengthened by the fact that Braque said also, "Art soars, science provides crutches" and defined science as "acquired power of repetition." But whether or not Braque's epigrams favor art over science, what is important is this: no matter how one interprets them, they remind us of the tension between these different ways of knowing [2].

If art upsets, it does so through its manifold views of the world. Art answers universal questions, but the reply varies from artist to artist. One artist's vision reassures one particular viewer, but eludes or disturbs another; a viewer finds one artist consoling, another upsetting. Science responds to the great questions as well, but does so by seeking singular and confirmed answers, which reassure in a way that art does not. Physicists now believe that a universal theory, a Theory of Everything, is within reach. And whether interrogating the origins of the universe or the processes of life, the driving hope is always that a response will emerge that satisfies most or all scientists. That mutual understanding must also be tested against reality by experiment. Such "correct" answers comfort us all, scientists and non-scientists; but consensus is no central aim of art, except in art artificially imposed, as under Nazism and Communism.

Perhaps exactly that difference between individual free response and laborious seeking of agreement is why, for Braque, "Art soars, science provides crutches."

It is not only art that liberates the spirit, and not only science that takes halting steps; but if there is any truth in Braque's epigrams, can art and science ever cast light on each other? Few human activities would seem to differ more in their intellectual bases, in the training and temperaments of their practitioners. Yet mutual illumination, I believe, is possible. Each—science and art, artist and scientist—includes more of the other than we usually recognize. But apparent similarities must be explored with care precisely because the goal is so alluring, so easy to applaud. Science does appear in art, artistic principles do enter science; but there are also cautionary tales where parallels have been too carelessly drawn.

One valid link between the areas comes from the study of human vision. Whether interpreted as actual seeing, or the imaginative use of the mind's eye, vision is too vital a human function not to appear in both art and science, where it sometimes forms a meaningful bridge. That was already true in the 18th century, when a great deal was known about light and vision. Even much earlier, in 1637, Rene Descartes had correctly shown that light rays were focused by the lens at the front of the eye onto the retina at its rear. Later in that century, Isaac Newton broke white light into colors, and John Locke contemplated human visual perception. And by the 18th century, two crucial visual properties were known and somewhat understood.

Accommodation, the ability of the eye to clearly view both near and distant objects, was one of these properties. Then as now, this was associated with changes in the focal length of the eye's lens, achieved by muscles that alter its shape and therefore its optical properties. The change depends on the color of the light as well, so while accommodation allows the eye to track from near to far, it also tends to make red objects appear closer than blue ones. The other property was acuity, the ability of the eye to resolve fine detail. The science of the time could not examine the microscopic structure of the retina, which ultimately determines resolution. It did know, however, that the greatest acuity occurs at a spot near the center of the retina, called the fovea; and that during the act of seeing, the eyes move to aim the fovea at different parts of the scene, to form a detailed image.

The art historian Michael Baxandall notes that this knowledge of the visual process permeated the intellectual air of the 18th century and was readily available to artists. He points to the work of Jean-Baptiste-Siméon Chardin, who enhanced French genre and still-life painting, as showing just that sense of a shifting point of best vision, as well as the act of accommodation. Chardin's *A Lady Taking Tea* of 1735 is a seemingly straightforward view of a woman seated at a table, gazing down at her teacup as she stirs its contents. But on close examination, some portions of the picture are seen to be sharply painted and highlighted, while others are fuzzy and dim. The table seems to advance toward the viewer because of its startling red, while the wall of the room retreats. Baxandall calls the painting ". . .an enacted record of attention which we ourselves. . .summarily re-enact. . .it has foci, privileged points of fixation. . .and curiosity about what it does not succeed in knowing," an apt description of how we see, a physiological process, and a placing of awareness requiring eye and brain together [3].

Later the scientific details of interrupted seeing became clearer. In 1878, the French physician L. E. Javal, studying how people read, found that the seeing eye constantly darts over a scene in short movements. These came to be called saccades, from the French for "twitch." They bring the fovea, with its high acuity, to bear on different elements of the scene. We are ordinarily unconscious of this changing gaze, which we now know flicks from position to position in a few milliseconds, remains at rest for a fifth of a second, then darts again. One of the earliest questions about saccades asked how these discontinuous glimpses take in a scene.

In the 1930's, the psychologist G. T. Buswell addressed the question through art, when he examined how people look at pictures. Buswell arranged a narrow beam of light to be reflected from the eye of a viewer. As the viewer altered the direction of his gaze, his eye acted like a small mirror whose shifts changed the direction of the reflected ray. Buswell captured the moving beam of light on photographic film, recording where the line of vision lay at any given moment. The experimental subjects were shown works by artists from Rockwell Kent to Georges Seurat, Winslow Homer to Marcel Duchamp, and encompassing scenes from Hokusai's famous wood-cut *Great Wave Off the Coast of Kanagawa* to cathedrals and contemporary advertising displays. The data showed that the eye does not shuttle regularly over a scene like a search team quartering an area, but follows intricate patterns of choice that emphasize areas of high interest and pay scant attention to the rest. The intricate mix of psychological attentiveness, neural control, and muscular versatility that appears in saccades is still studied by more sophisticated methods, but it was Buswell's use of art that established the complexities of our apparently disjointed vision.

In Chardin's art and Buswell's research, the connections between scientific and artistic examination of human vision seem natural and unforced. When the vision is of a more abstract sort, it is easy to fall into forced and erroneous connections. That happened with the Cubism of Braque and Picasso, which became mistakenly associated with Einstein's Theory of Relativity. Some Cubist paintings, such as Picasso's *Portrait of Daniel-Henry Kahnweiler* and *Portrait of Ambroise Vollard,* both from 1910, seem to show all sides of the subject at once, through multifaceted prism-like segments of the image. The artist's vision appears to encompass a three-dimensional subject more globally than does ordinary sight. That enlarged vision became connected to the 19th-century idea of a spatial fourth dimension, which then seemed to associate itself with Einstein's Theory of Relativity.

The reality we know contains only three spatial dimensions; but we can speak of a fourth dimension by making analogies to simpler geometries. In 1827, the mathematician A. F. Möbius—who also invented the Möbius band, a strip of paper joined end-to-end with a half-turn that gives it only a single surface—noted the power of greater dimensionality. Consider the flat soles of your shoes. As long as they are confined to the two-dimensional floor, it is impossible to make the right one look identical to the left, no matter how you turn it. But one can be made to exactly match the other if you lift it into the third dimension and rotate it. An added dimension gives the versatility of an extra axis in space. Similarly, noted Möbius, right-and left-handed shoes, gloves, or any other paired three-dimensional bodies could be turned into each other, if there existed a fourth dimension perpendicular to our present three.

This idea of higher spatial dimensions traced an intriguing thread through the century, partly through mathematical interest and partly because of popular fascination with its magical properties. The 1891 fantasy *Flatland: A Romance of Many Dimensions,* by E. A. Abbot, put the basic concepts into charming form. The inhabitants and structures of Flatland are mere surface patterns without thickness. None of Flatland's creatures can ever gaze over a neighbor's wall, whereas their homes and treasure chests, their very internal organs, are open to our three-dimensional sight. If we were to cruelly push a finger through that world, its citizens would see only a line whose length changed as the finger moved up and down, their edge-on view of the varying cross section that intersected their particular reality.

We cannot be smug about our own three-dimensionality, because what we can do to the Flatlanders, four-dimensional beings can do to us. Such a creature could simultaneously see our every part with perfect ease. It could extend a tentacle to rummage among our insides. We would see only an inexplicably writhing three-dimensional shape, just as the Flatlander interprets the probing finger as a mysteriously changing line. It only added to the mystique of higher dimensions that mathematicians could in fact derive the three-dimensional projections of four-dimensional objects like the tesseract, the extended version of the cube.

As Linda Dalrymple Henderson shows in *The Fourth Dimension and Non-Euclidean Geometry in Art,* these ideas had become an artistic staple by the early 20th century, making it natural to describe Cubist works as if they displayed a four-dimensional view of the world. She notes, in fact, that the triangular facets of *Portrait of Ambroise Vollard* bear an uncanny resemblance to mathematically accurate representations of certain four-dimensional solids. But although the fourth dimension had a valid mathematical interpretation, it had never attained physical reality, until it seemingly did so in Einstein's Special Theory of Relativity. That came in 1905, nearly at the same time that Cubism was being created [4].

Einstein's marvelous insight was that time and space are not immutable, as classical physics had assumed; instead, they change according to the speed of the observer, with enormous alterations near the speed of light. The distance between two locations in space, or the time between two events, was no longer fixed. But Einstein found a new combination of time and space that remained unchanged. He joined time with the three spatial dimensions to make a single mathematical entity, the "proper" spacetime interval, which would remain the same for any viewer. That in turn suggested to the mathematician Hermann Minkowski that the universe be treated as a four-dimensional spacetime continuum.

Although Minkowski's world is tantalizingly like the old idea of the fourth dimension, it is not the same. It is only one of the many examples where scientists embrace dimensions beyond the ordinary three for mathematical convenience, but without implying that they are real, or introduce spatial strangeness. Physicists sometimes choose, for instance, to describe the motion of an object in six dimensions: three defining its location east–west, north–south, and up–down; and three defining its velocity in each of these directions. Similarly, the fourth dimension of time in Einstein's universe is not the fourth dimension of space that would allow a left-handed glove to become right-handed. Yet those different meanings of "fourth dimension" have led some to confound the invention of Cubism with the invention of relativity, and even to argue that the artistic insight preceded the scientific one. Einstein himself rejected the association, saying, "This new artistic 'language' has nothing in common with the Theory of Relativity" [5].

Such confusions should not obscure the many valid ways in which physical science appears in art. The radical German artist Joseph Beuys, who died in 1986 after a vast outpouring of varied works, was intensely concerned with the meaning of materials. His charged experiences in World War II gave felt and animal fat a particular emotional weight. He also understood the physical properties of substances. He related his sculptural use of fat and wax to processes that change amorphous substances into geometric and crystalline forms. Copper was for him a transmitter of intangible energy, in metaphorical agreement with the fact that it is nearly the best metallic conductor of electricity. Electric current has metaphorical value as well for the contemporary New York artist Eve Andrée Laramée, whereas the late Kansas City artist Dale Eldred used optical science to turn light into pure color and into seemingly tangible structural elements in his installations.

Apart from specific cases, there is a universal crossroads where art and science may intersect. Some pictures radiate beauty through balance and harmony; some scientific theories show beauty in their intellectual rightness, the property scientists and mathematicians call "elegance." Elegance may mean concise completeness, a rich simplicity that explains much with a few ideas or symbols. James Clerk Maxwell, the greatest mathematical physicist of the 19th century, left memorable achievements in statistical physics, planetary studies, and the theory of color; but his overarching feat is the set of four brief equations that describe nearly everything we know about electricity and magnetism. With a handful of mathematical symbols, they summarize the repulsion of like electrical charges and the attraction of unlike ones; the shape of the curved web that iron filings form when sprinkled near a magnet; the reason that megawatts of electrical power can be made by enormous rotating generators; radio and television waves; and light waves and the speed of light. They omit only light as quantum particle, for the photon was found long after Maxwell's time. Still, this compact theory remains central to physics and to our technological world.

Elegance can also mean a sure-footed route to an answer picked through rugged intellectual terrain. In the early 20th century, physicists began unwillingly to conclude that nature had a hidden complexity, for there was mounting evidence that light somehow appeared in the form of waves and particles at the same time. The French theoretical physicist Louis de Broglie cut through the confusion and went straight to the heart of the matter. He derived a simple equation whose three symbols for the first time linked wave to particle, and which also correctly predicted that particulate matter such as electrons had wave-like features. De Broglie's result did not fully explain the wave–particle duality, which remains enigmatic even in later theories, even today; nevertheless, his unadorned insight achieves elegance through its concise power, like an artist's quick sketch that captures the essence of the subject before the full picture is painted.

Scientists agree that one theory may be more elegant than another; they disagree, however, as to whether beauty is necessary in scientific thinking, or even desirable. One might think that mixing aesthetics into science can never hurt, but beauty sometimes upsets truth. In his geocentric theory of the solar system, the astronomer Claudius Ptolemy of Alexandria assumed around 100 AD that the planets move in perfect circles. Those supremely symmetric rings owe something to a vision of ideal beauty that dates back to Plato and later became connected to a kind of religious aesthetics. But the supposed perfection of form obscured things as they really are.

It was not until 1609 that Johannes Kepler, having examined the astronomical data, announced that planetary orbits are ellipses, not pure circles. That observation eventually led to radical revisions in our views of the solar system and the cosmos.

Yet the same aesthetic elements that misled Ptolemy may guide other scientific quests. Symmetry is a true feature of nature, appearing in the spherical shapes of microscopic protozoans, in the bilateral symmetry of higher animals, in solid crystals. There is scientific meaning in such inherent balance, and the search for balance has stimulated explorers of nature. In the early 19th century, the German scientist Johannes Ritter noted that certain contrasting physical properties appear in linked pairs. Magnetic north and south poles, for instance, are opposite in physical effect, but appear together in any magnet. Ritter knew that the astronomer William Herschel had in 1800 discovered invisible infrared radiation by searching past the last trace of red light in the visible spectrum. Applying his principle of linked opposites, Ritter reasoned that unseen radiation should also lie beyond the other end of the visible spectrum, where we see the color violet. In 1801 he found that energy, now called ultraviolet light.

Although elements of beauty appear in science, the real question is whether an overriding aesthetic ideal should determine scientific thinking. At least one scientific titan, the British theoretical physicist Paul Adrien Maurice Dirac, believed that it should. Dirac earned a Nobel Prize in 1933 for his extraordinary theory that combined quantum mechanics with relativity. The theory explained some of the puzzling properties of electrons and other elementary particles. And it made a startling prediction that has been confirmed by experiment, that elementary particles have conjugate entities called antiparticles—the birth of antimatter. Dirac was led to the equation that gave these wondrous results by his belief that only beautiful mathematical descriptions of nature can be correct. "It is more important to have beauty in one's equations than to have them fit experiment," he said. I myself am an experimental physicist, and this statement makes me acutely uncomfortable; but it is undeniable that Dirac's sense of mathematical aesthetics led him to his most original ideas [6].

Albert Einstein might have disagreed about the value of beautiful equations, at least when he developed his theory of gravity in 1915,

General Relativity. Gravity is the only great force of nature that is not fully understood, although it has been contemplated for 300 years and more. In 1686, Isaac Newton published his law of universal gravitation, which gives the gravitational pull of one body on another, whether it be the earth on a stone, or the sun on the earth. The law was a triumphant success, explaining Kepler's elliptical orbits, the motion of the moon, the behavior of falling bodies, and the ocean tides; yet it is embodied in a mathematical relationship so simple that beginning students of physics can apply it.

This law remained unchanged until Einstein created a new vision of gravity. General Relativity does away with the notion of a force acting between bodies; it ascribes gravity instead to the bending of the spacetime continuum around a massive body, as a stretched tarpaulin is deformed when it supports a bowling ball. This abstract warping of spacetime is nearly impossible to visualize; fittingly, it is described by formulas so complex that Einstein had to invent a new mathematical notation to express them. He replaced Newton's single equation with a set of ten interlinked relationships, each of the type called a second-order partial differential equation. These are difficult enough to solve individually and are still not fully understood in their totality.

Newton's relation is lucidly elegant; Einstein's, dauntingly difficult. Einstein was once reproached for this awkward formulation. His reply throws issues of truth and beauty into sharp relief: "If you are out to describe the truth," he said, "leave elegance to the tailor." Einstein's ungraceful equations carry greater truth than Newton's result, for they explain aspects of planetary motion that Newton's theory does not; predict black holes, those strange cosmic features that suck in light, one of which has recently been found; and even dare calculate whether the cosmos is shrinking or expanding. In its fuller description of reality, its greater predictive power, General Relativity outstrips the Newtonian formulation. And for all its obscure mathematics, the theory is conceptually elegant; by connecting gravity to an innate cosmic geometry, General Relativity avoids certain pitfalls that arise in Newton's idea of gravitational force [7].

To speak of Newton's gravitational force, Einstein's geometric universe, Dirac's beautiful mathematics, is to realize that each thinker followed his own path to truth. The scientific stamp of approval, the final reassuring weight of the accepted answer, can only validate the insights of individual scientists. And just like so many artists, each scientist follows his own muse, to create and develop ideas within his own style. If personal artistic freedom is the foundation of true art, scientific freedom is necessary for science. This is their great common feature: each transforms passionate individual questioning into answers.

Perhaps, then, the truer version of Braque's *mot* is "Artists upset, scientists reassure." The white-coated medical scientist, the dispassionate astronomer, represents truth as defined by the acceptance of many scientists. To clinch an argument, one need only utter, "Scientists say;" never, however, have I heard the claim, "Artists say." That is partly because artists do not speak with a single voice, but there is more. Compared to scientists, artists are seen as intuitive rather than rational, off-beat rather than conventional. They are thought to untidily mix life with work, whereas scientists keep their research coolly distant from ordinary human concerns. That objective distance makes scientific pronouncements reassuring. It is why the fictional figure of the evil "mad scientist" is so dramatic; he stands in contrast to the predictable rationality of his colleagues. But the evil "mad artist" is not much seen in fiction, for pure rationality is less central in art. Artists also employ subjective, non-intellectual modes of understanding, as they did explicitly in the movements of Dadaism and surrealism that flourished after cubism.

There is no scientific equivalent of Dadaism, because scientists want to objectify and minimize the irrationality of the world, not embrace it. They cannot make fully valid science out of subjective response and personal expression, as artists can weave art. Sometimes extreme personal acts lift an artist to the status of romantic icon, but that does not happen to scientists. The Nobel Prize-winning British biologist Peter Medawar notes that the private lives of scientists have no special impact on their work. In a direct comparison to Vincent van Gogh, he writes: "If a scientist were to cut off his ear, no one would interpret such an action as evidence of an unhappy torment of creativity; nor will a scientist be excused any *bizarrerie*, however extravagant, on the grounds that he is a scientist, however brilliant." But there is a compensation: as the personal lives of scientists are less colorful and less deeply embedded in their work, their pronouncements are more reassuring [8].

This, it seems to me, is the truest meaning of "art upsets, science reassures." In showing that rational thought leads to understanding, science comforts us in a vast universe. In its willingness to enfold the unfathomable, art can be profoundly troubling; yet its answers give meaning even when rational understanding fails. Both endeavors grow from a deep human need to find our place in an indifferent and enigmatic cosmos, a search that may never reach its goal. In a final burst of mutual illumination, artist Georges Braque and scientist Albert Einstein both show that they understand this power of the inexplicable. "There is only one thing in art that has value," said Braque, "that which one cannot explain." And, said Einstein, "The most beautiful thing we can experience is the mysterious. It is the source of all true art and science"

Bibliography

Alan L. Mackay, *A Dictionary of Scientific Quotations* (Institute of Physics Publishing, Bristol, 1992).

Edwin Abbott, *Flatland: A Romance of Many Dimensions* (Barnes and Noble, New York, 1963).

Georges Braque, *Cahier de Georges Braque* 1917—1947 (Might, Paris, 1948).

G. T. Boswell, *How People Look at Pictures: A Study of the Psychology of Perception in Art* (University of Chicago Press, Chicago, 1935).

Linda Dalrymple Henderson, *The Fourth Dimension and Non-Euclidean Geometry in Art* (Princeton University Press, Princeton, 1983).

Michael Baxandall, *Patterns of Intention: On the Historical Explanation of Pictures* (Yale University Press, New Haven, 1985).

Peter Medawar, *Advice to a Young Scientist* (Basic Books, New York, 1979).

References

1. "I do not search, I find" is cited in Mackay, 194; "Don't talk to the driver" in Baxandall, 41.
2. "Art is meant to upset...," Braque, 10; "Art soars...," Braque, 19; "Science is acquired power...," Braque, 62.
3. Baxandall, 102.
4. The comments about *Portrait of Ambrose Vollard* appear in Henderson, 58.

5. Henderson, 356, quoting Paul M. Laporte, "Cubism and Relativity," *Art Journal* **XXV**, (Spring, 1966, 246).

6. "It is more important...," Mackay, 74.

7. "If you are out to describe the truth..." is ascribed to Einstein in Mackay, 81. Einstein may have been repeating the comment; I have seen it ascribed to at least one other physicist, expressed in German where it plays on *Schonheit* and *Schneider,* "beauty" and "tailor."

8. "If a scientist...," Medawar, 40.

9. "There is only one thing...," Braque, 13; "The most beautiful thing..." John Bartlett, *Familiar Quotations,* (Little, Brown and Company, Boston, 1992).

Hubs, Struts, and Aesthetics

Years ago, when I began the hard work of engineering school in New York City, I still found time to haunt the Museum of Modern Art (MOMA) on 53rd Street in Manhattan. While I grappled with calculus and physics, I also encountered the intensity of Vincent van Gogh's *Starry Night*, the strangeness of Rene Magritte's *Empire of Light*, the delicacy of Claude Monet's *Water Lilies*. Those powerful works spoke to my aesthetic side, but they were not all that MOMA had to offer. On its top floor I encountered another kind of modern art, the art of made objects. Such works spoke to my aesthetic side and my rational side alike, as I learned to appreciate the balanced simplicity of a record turntable, the rightness of form in a laminated wooden airplane propeller, the spatial intelligence contained in a chess set that nested together to exactly fill its storage box.

My career was to turn toward physics, not engineering, but no matter which path I followed, the objects on MOMA's top floor would have lighted the way. They embodied in metal, wood, and plastic an idea that underlies the best in science and technology—the notion of elegance, of concise and beautiful solutions to problems, an idea expressed equally well in powerful mathematical equations, incisive physical theories, and marvelous devices. And because those objects of elegant design united aesthetics with function, they taught me to see with a dual vision that blends satisfaction in knowing how the world works and appreciation of its beauties.

That vision still lets me perceive that our daily world of manufactured objects is full of elegant design. Every city street, for example, houses a fascinating gallery of techno-art—the works of machine aesthetic that grace the wheels of every passing automobile. Within the functional geometry of each wheel, a circular rim joined to a hub fastened to an axle, designers are free to decorate in bewildering variety.

That was not possible for the first wheels, invented in Mesopotamia in the fourth millennium BCE. Those solid disks pieced together from wooden planks were not easily worked into decorative shapes. The spoked form that appeared a millennium or two later in war chariots, and then in vehicles from farm wagons to royal coaches, was a better wheel and offered more design choices; spokes could assume any number of ornate shapes and color schemes. Later still, spokes of iron or wood showed up on early steam engines, and then on automobiles and bicycles. Some evolved into the thin struts of wire wheels, which combined flexible strength and low weight with airy charm. Spokes survive today, but now are only one among many styles of automotive wheels.

Indeed, wheel design, including decorative hubcaps and covers, expresses all sorts of ideas, even class distinctions. On a small truck or commercial van, each wheel is fastened to a naked hub by visible, perfectly plain nuts—usually five of them—with a simple rib radiating from each nut. Move up from a utilitarian vehicle to a sedan and you see a wheel prettified, but with its mounting nuts still accessible. In some cases, the hub and nuts lurk under a removable cover, as if to shield innocent eyes from the reality of machinery. In others, nuts are exposed but are integrated into a wheel with ribs, struts, slots, or open ports, whose number seems to rise with the price of the sedan.

Some wheel designs resonate, intentionally or otherwise, with ancient artistic symmetries: distinctive designs with 14 or 18 ribs seem to echo the circular stained-glass rose window (also called a wheel window) at Notre-Dame Cathedral in Paris—a window with 16 lobes—or the stylized chrysanthemums in Japanese art, which have 14 or 18 identical petals.

Car designers take wheels very seriously. One automaker I recently approached treats its designs as top secret, refusing to discuss even past wheel designs. Over at Toyota U.S.A., however, I found a knowledgeable guru of wheels. Don Brown, the company's product planning manager, points out that wheels were once created solely by engineers but that a growing emphasis on aesthetics has drawn designers into the mix.

Automakers worry about the look of wheels because customers notice them—if not overtly, then as a subliminal part of a car's total presentation. European car companies like Ferrari have known this all along. Adding the wrong wheels, says Brown, can be as jarring as "wearing the wrong shoes with a tuxedo." Wheels with five plain, chunky ribs match a pickup truck the way work boots go with jeans. The wheels for the first model of the Lexus sedan, in contrast, went through three design cycles before they looked fittingly luxurious.

In designing some parts of a car, automakers can choose whether to highlight style over utility. For instance, every internal-combustion automobile has plumbing to carry exhaust gases from the engine to the open air. Some classic cars of the 1930s sported spectacular chrome-plated exhaust lines that looped from hood to fenders. But the usual design solution has been to decently bury the entire exhaust system under the car; the only concession to aesthetics might be to beautify the visible tip of the exhaust pipe.

Wheels, however, must work well and look good. This combination proved elusive in the days when wheels were made exclusively of steel. What was handsome might also be heavy, hampering performance and fuel economy. Now designers can use easily worked aluminum, metal alloys, and plastics to make wheels that are not only strong and lightweight but also attractive. Still, a host of engineering constraints remain. A designer might imagine a look based on, say, four mounting nuts rather than five; but engineers tend to prefer five for strength and reliability. Or the designer might want to add three-dimensional elements; but if they protrude too far, they could jam in the tight clearances of a car wash.

Sometimes the mixture of engineering and aesthetics rises above mere coexistence. That happens in sports cars, where the wheels—like the rest of the car—say "this vehicle can perform" and hint at the underlying machinery. The wheels are generally businesslike, with five exposed nuts and five ribs that are sturdy, but shaped to suggest speed—for instance, with a dynamic, tornado-like swirl. The ribs frame large openings to cool the brakes, which remain largely exposed to view. This design might be undesirable in a luxury sedan, but in the high-performance Toyota Supra the physical arrangement and finish of the brake mechanism were carefully chosen to complement the rest of the car.

Even the open ports between the ribs can unite form and function. The wheels on my Nissan 300ZX sports car display subtly asymmetric openings, each defined by a compound arc shaped like a boat's sail taut against the wind. This lovely curve, it turns out, is designed to scoop air from beneath the car, past the brakes to cool them, and out through the wheels.

At rest, the outdoor gallery of wheels is an attractive assortment of still lifes. But seen in dynamic display, with wheels spinning, brakes heating, and air rushing, it represents the elegant blend that defines true techno-art. A designer of wheels might take professional satisfaction in these stylish answers to engineering problems. For me, the blend suggests an answer to another, larger question I am sometimes asked: Is my enjoyment of good design, or of beauty, diminished if I think of the underlying physical reality? Does a handsome manufactured object, or a distant landscape, become less compelling if I consider the details of metallurgical processing, of geological formations?

The answer, of course, is that the time I spent studying calculus and van Gogh, physics and the curve of a propeller, only enhanced my vision by focusing it through a scientific lens. Anyone with scientific training has the power to look both at and through the surface of the world, the power to merge visceral reaction and cool analysis into heightened response.

Inspirational Realism: Chesley Bonestell and Astronomical Art

Chesley Bonestell was born in San Francisco in 1888 and came to astronomical art early, inspired as a teenager after seeing Saturn through a telescope at the Lick Observatory near San Jose. However, he was trained as an architect and contributed both to a famous West Coast structure, the Golden Gate Bridge, and to an equally famous East Coast structure, New York City's Chrysler Building. In 1938, he became a special effects artist who painted background scenes for films. In this way he also contributed to a famous imaginary structure, Charles Foster Kane's castle "Xanadu" in *Citizen Kane* (1941), considered by many the all-time greatest American film (Miller 1994).

Bonestell's experience in film heightened his awareness of the importance of camera angle and point of view as directed toward a scene. Like a cinematographer, he incorporated this sensibility into his astronomical art, which added to its immediacy along with his hard-edged, almost hyper-realistic style.

The scientific validity of Bonestell's work was confirmed in the 1950s when he began collaborating with space scientists, especially Wernher von Braun, a leading rocket scientist who had worked on Nazi V2 rockets during World War II and then became instrumental in the U. S. space program. Von Braun greatly esteemed Bonestell's devotion to precise depiction, saying that Bonestell's pictures

> present the most accurate portrayal of those faraway heavenly bodies that modern science can offer . . . In my many years of association with Chesley I have learned to respect, nay, *fear*, this wonderful artist's obsession with perfection. (Schuetz 1999, p. xxii; Durant and Miller 1983, p. 9–10).

In a series of books, magazine covers, and articles, and background art for films, Bonestell's images in the 1950s of what awaited us when we left Earth, along with images of the spacecraft and equipment we would use, virtually defined "space" for the public. His output is credited with helping create an atmosphere that favored and supported the idea of human exploration of space.

Bonestell's use of unusual viewpoints is illustrated in his images of the planets of our solar system, for instance his views of Mars and Jupiter from their respective satellites. In his image of Jupiter, Bonestell was ahead of his time in putting the small satellite Europa in the foreground rather than the infinitely more imposing Jupiter; scientists now believe that Europa is highly significant as one of the few bodies in the solar system with liquid water.

When it came to depicting the hardware that would carry humanity into space, Bonestell was strongly influenced by von Braun, who carefully worked out designs for everything from multi-stage rockets that could lift off the Earth to craft designed to land on the Moon and Mars. Bonestell's drawing of a lunar lander blasting its retro rockets as it descends onto the Moon's surface puts life and color into von Braun's detailed engineering drawing. Echoes of this design, with its lack of streamlining for an airless environment, its crew compartment above, and its landing struts suitable for a rough surface, can be seen in NASA's Apollo 11 Lunar Module that landed Neil Armstrong and Buzz Aldrin on the Moon in 1969, though von Braun had designed a much bigger craft.

One of the most influential results of the von Braun–Bonestell collaboration was a series of articles published in several issues of *Collier's Magazine*, from March 1952 to April 1954, that traced the steps of developing a space program and showed how we could put humans on the Moon and Mars. Bonestell's lunar lander graced the cover of one of these issues, and one of his interior illustrations for the series again shows his use of dramatic viewpoints, as we look down at a spacecraft under assembly in Earth orbit with a clear view of the Earth below. It also demonstrates the engineering detail that von Braun provided and that Bonestell expressed so well in his art. The winged spacecraft is von Braun's design. Remarkably, the image also shows a telescope floating in space that is amazingly similar to NASA's Hubble Space Telescope launched in 1990, nearly four decades after this image appeared.

In addition to his efforts for books and magazines, Bonestell continued his work for films and television through the 1950s. Acting as technical adviser, he contributed to the look of several classic science-fiction movies that also shaped public perceptions of space and of technology. These include *Destination Moon* (1950), based on a story by Robert Heinlein and featuring realistic depictions of a spaceship and of the Moon's surface; *When Worlds Collide* (1951), about the destruction of the Earth through a collision with an incoming space object; *The War of the Worlds* (1953), based on H. G. Wells' 1898 book of the same title, about invading Martians; and the television series *Men into Space* (1959). He also painted striking cover art for the two leading science-fiction magazines of the era, *Astounding* (later *Analog*) *Science Fiction* and *The Magazine of Fantasy and Science Fiction.*

Bonestell's role in making space real to the public and in inspiring young people to enter into space research, astronomy, or other science and engineering careers, was so widely recognized and respected that both a crater on Mars and an asteroid were named after him. On the occasion of naming that asteroid "3129 Bonestell," astronomer Carl Sagan perfectly summed up Bonestell's great contributions when he said, "It is only fitting that we give back a world to Bonestell, who has given us so many" (Miller and Durant 2001, p. 107)—which could also serve as Bonestell's epitaph after his death in 1986.

Bibliography

Durant, F. C. and R. Miller. *Worlds Beyond: The Art of Chesley Bonestell* (Norfolk, VA: Donning, 1983).

Ley, Willy *The Conquest of Space,* paintings by Chesley Bonestell (New York: Viking Press, 1950).

Ley, Willy, and Wernher von Braun, *The Exploration of Mars,* paintings by Chesley Bonestell (New York: Viking Press, 1956).

Miller, R. "Chesley Bonestell's Astronomical Visions," Scientific American, **270**, 76 - 81, May 1994.

Miller, R. and F. C. Durant, *The Art of Chesley Bonestell* (London: Paper Tiger, 2001).

Schuetz, M. H. *A Chesley Bonestell Space Art Chronology* (Parkland, FL: Universal Publishers, 1999).

von Braun , Wernher, Fred L. Whipple and Willy Ley, *Conquest of the Moon,* illustrated by Chesley Bonestell, Fred Freeman and Rolf Klep, edited by Cornelius Ryan (New York: Viking Press, 1953).

Art, Physics, and Revolution

It was not the kind of newspaper story you would ever associate with a noted artist: datelined Mexico City, October 5, 1940, the *New York Times* headline read, "TROTSKY SUSPECT CAUGHT: Siqueiros, Painter, Wanted for Abortive Attempt on Exile."

"Siqueiros, painter" was David Alfaro Siqueiros, who with Diego Rivera and José Clemente Orozco formed "Los tres grandes," the leaders of the Mexican mural artistic and social movement of the 1920s to 1940s. "Trotsky," Leon Trotsky, was the Marxist theorist and founder of the Red Army who substantially helped Vladimir Lenin complete the communist takeover of Russia in 1917.

After Lenin's death, Trotsky lost the battle for supremacy in the Soviet Union to Joseph Stalin and fled into exile in Mexico City. Siqueiros, an ardent communist who supported Stalin, tried to assassinate Trotsky there by leading a murder squad that fired hundreds of machine-gun bullets into his bedroom. Siqueiros was imprisoned for this attack, though it did not kill Trotsky, but was released when Trotsky was bloodily dispatched by a different assailant with an icepick.

That was one chapter in Siqueiros' impassioned life, a life that mixed art and revolution yet also connected art to science and technology. Siqueiros' life spanned cultures as well, for his art had widespread effects. At an experimental workshop he established in New York in 1936, he demonstrated "accidental painting" and other new artistic methods, which he felt essential to the development of a true revolutionary art. As one workshop attendee wrote, "We sprayed [paint]. . .used it in thin glazes or built it up into thick globs. . .we poured it, dripped it, spattered it, hurled it at the picture surface." Among the participating artists was the 24-year-old Jackson Pollock, who went on to develop his own unconventional but now famous "drip" technique of dropping paint onto a horizontal canvas, thereby changing modern art and helping to create Abstract Expressionism.

"Accidental painting" implies randomness. As discovered in 2012 by art historian Sandra Zetina and physicist Roberto Zenit of the National Autonomous University of Mexico (UNAM), one of Siqueiros' methods of manipulating paint indeed gives unpredictable patterns as it arises from a specific effect in the physics of fluids, the Rayleigh–Taylor (RT) instability. On the other hand, in 2011, Andrzej Herczyński and Claude Cernuschi of Boston College, and L. Mahadevan of Harvard, analyzed the fluid dynamics of Pollock's approach to show how it allowed him to guide a stream of paint to canvas in predictable ways.

These analyses provide a thread joining the two artists and their legacies, and raise intriguing questions about the scientific investigation of art and about randomness versus artist's intention— not that painters need scientific insight to make art. They use their empirical knowledge of the malleable properties of paint to mix colors and apply the result with brush or tool to produce the desired shade and texture on a surface.

But though cave artists were using crude forms of paint thousands of years ago, and oil paint has been the dominant fine art medium since the 15th century, we still do not fully know why paints behave as they do—yet deeper understanding would be desirable. Think of photographic art: a person can learn to operate a camera and produce good photographs; but a person who also grasps the basis of the art and its tools—optics, the properties of lenses and light sensors, image manipulation, and so on—can produce better ones. Similarly, the painter who fully understands the behavior of paint and how to exploit it can make better art.

Paints, though, are hard to describe scientifically because they are complicated. The categories for types of matter include solids, liquids, and gases, with the latter two often combined as "fluids." But these clear divisions do not apply neatly to paints, which consist of fine solid grains of pigment (such as the compound cadmium sulfide, the yellow in cadmium yellow) suspended in a liquid carrier (usually linseed oil in oil paints). This combination is crucial to the art of painting, for the liquid carrier allows the paint to be applied as desired; then the carrier evaporates, leaving behind layers of color to form the work of art.

With their mixed solid–liquid nature, paints are "complex" fluids, as opposed to "simple" ones like pure water. But even "simple" fluids are not simple. Though water seems utterly lucid and straightforward, its motion is difficult to analyze. It flows and swirls in non-linear ways, yielding unexpected results for small changes and supporting seemingly random turbulent eddies. The great 18th-century Swiss mathematician Leonhard Euler derived an equation for fluid motion, but to his embarrassment, when he used it to design a decorative fountain for Frederick the Great of Prussia, the fountain did not work. Scientists have since produced the correct equation for fluid behavior but it is extremely challenging to solve. Even with modern computers and experimental techniques, fluid dynamics remains a difficult research area in physics, all the more so for a complex fluid like paint.

Still, science can explain some important fluid effects, such as the RT instability, named after the British physicists Lord Rayleigh and Sir Geoffrey Taylor. Rayleigh won a Nobel Prize in 1904, but is best known for answering an age-old children's question: why is the sky blue? He found the answer in 1871: the molecules in the atmosphere deflect or scatter sideways the blue part of the rainbow of color contained in sunlight but let the other shades pass straight through. The result is the canopy of blue we see when we look up at the heavens.

Then in 1880, still looking heavenward, Rayleigh studied how clouds form when two fluids, water vapor and air, interact. He found that when a fluid supports a second denser fluid atop itself, the situation is unstable, like a big rock precariously balanced on a small one. Any tiny disturbance at the interface is pulled by gravity into the less dense fluid and quickly grows. The fluids mix turbulently and form complex finger-like intrusions into each other. In 1950, Taylor extended the scope of the effect by noting that the instability can occur when any force, not just gravity, pushes one fluid into another.

The RT instability is widespread in nature. It enters into certain types of supernova stars and can be seen in huge interstellar gas clouds such as the Crab Nebula; it played a role in the formation of the Earth's core and affects how the oceans transport heat, therefore influencing the Earth's climate; and it has biological significance. In human use, RT instability affects the detonation of atomic bombs, as Taylor noted during World War II, and appears in today's attempts to produce clean and limitless fusion power from hydrogen.

RT instability also appears in Siqueiros' art as Zetina and Zenit found at UNAM. Siqueiros did not consciously apply the laws of fluid dynamics, but scientific analysis of his work is satisfyingly consistent with his principles as a revolutionary artist interested in technology for artistic purposes and also to create a new society. Like Lenin,

Siqueiros believed in the power of technology to build a communist future in Russia or his own Mexico.

These threads united when Siqueiros, expelled from Mexico in 1932 for his radical politics, arrived in Los Angeles and discovered some useful products coming out of North American industrial technology. One was the spray gun, which he used to paint the murals *Street Meeting* and *Tropical America* on cement walls (he could not entirely give up his suspicions of capitalism, however; difficulties with the spray gun led him to write "I suffered many secret disillusions because of this fickle instrument invented by capitalist industry.") The spray gun was loaded with a paint called cellulose nitrate lacquer, which led to his exploration of RT instability. The paint, developed by another capitalist entity, the Ford Motor Company, did not require long periods of baking to dry and so allowed Ford's factories to maintain a brisk rate of production.

In 1936, as part of his New York Workshop, Siqueiros applied this paint to wood to create *Collective Suicide* (Museum of Modern Art, New York). This large painting (124 cm × 183 cm) depicts in horrific terms the 16th-century Spanish conquest of Mexico. It shows Spanish troops in armor killing or taking into slavery native Indians—some jumping from cliffs rather than submit—and destroying their culture. The work employs the radical techniques Siqueiros espoused, including elements seen in Pollock's later work, paint poured directly from the can or dripped from sticks onto the painting placed flat on the floor.

The painting displays RT instability as well, as shown in a video made by Zetina and Zenit at UNAM. It zooms in to show in close detail certain areas characterized by pleasingly intricate abstract patterns of intermingled organic-seeming blobs, swirls, and networks of color—black, white, gray, gold, brown, and more, in subtle variations and shadings. These areas come from Siqueiros' "accidental painting" technique, which he described in a letter:

> [When] pouring layers of paint, of different colors...they infiltrate into each other. [They] produce the most magical fantasies and forms that the human mind can imagine...similar to the geological formation of the earth, to the multicolored and multi-shaped seams of the mountains... similar to the synthesis, the very equivalence of the whole creation, of life.

The "fantasies and forms" came from RT instability, as Zetina and Zenit showed by their experiments. In one example, they poured black paint onto a horizontal surface, then poured a layer of white paint on top of the black. The top layer was denser than the bottom one, fulfilling the condition for RT instability. The video shows the result: in a hypnotically fascinating unfolding, black blobs magically appear, grow, and change shape and shading to mottle the white layer in an aesthetically satisfying arrangement that could not have been predicted in detail. The process stops when the paint dries, freezing the pattern induced by RT instability.

Similar patterns appear in trials Zetina and Zenit carried out with paints of different colors, densities, and viscosities. Zenit, the physicist, also showed that the observed behavior follows the appropriate mathematical theory for RT instability. And while the science helped elucidate the art, the art also advanced the science. When the instability completed itself and became fixed, that provided valuable new data.

Zetina and Zenit note that Siqueiros thought the paint itself created the abstract patterns in *Collective Suicide.* That is correct: once Siqueiros chose and layered the paints, the RT instability took over and produced a result outside his control. But though Jackson Pollock also used the fluid properties of paint, Zetina notes that the accidental nature of the RT instability is

> quite different from what Jackson Pollock described for his technique...
> [referring to one of his paintings] he says very clearly "this is not an
> accident. I'm controlling everything I'm doing" and the difference is
> evident in the results the two obtained.

Herczyński, Cernuschi, and Mahadevan (HCM) at Boston College and Harvard confirm that Pollock was right. Calling Pollock's approach a "drip" technique is a misnomer, they say, since his paintings do not typically display separate spots of pigment. Rather, works like *Autumn Rhythm* (1950, Museum of Modern Art, New York) consist of "a web of sinuous and undulating curves created by continuous filaments of paint" and should more properly be called "stream" painting. Films of Pollock at work support this reformulation. They show him moving around the canvas on the floor like a dancer as he repeatedly dips an implement (typically a hardened paint brush) into a can to load it with paint, then smoothly manipulates it to maintain a constant falling stream. HCM analyze the fluid dynamics of this technique to show that it provided Pollock with good control over the line of paint on the canvas.

Though Pollock's streaming was relatively controlled, Zetina, Zenit, and others believe that his early encounter with Siqueiros' accidental painting and radical artistic approach influenced Pollock's style. The exposure may have come even before 1936; according to Helen Harrison, director of the Pollock–Krasner House and Study Center in East Hampton, NY, Pollock knew about Siqueiros' methods as early as 1932. As discussed in her book *Jackson Pollock* (2014), she thinks these experiences were crucial to the development of Pollock's technique—though, as artists invariably do, he transmuted them according to his own aesthetic vision.

Siqueiros too might have pursued his own form of abstract art through accidental painting. But instead he created a paradox that Zenit notes: Siqueiros' "most well-known paintings are relatively traditional realist types," yet "he was a key element in modern art." As a devoted communist and in keeping with the thrust of Soviet art, Siqueiros chose to pursue representational rather than abstract art; nevertheless, he was quite sure that he had strongly affected Pollock. According to Zetina, later in life Siqueiros "showed his more experimental paintings [and said] look here, Pollock was there, I showed him, I was the man with the idea but I didn't follow it because of my convictions."

From today's vantage point, Siqueiros declaring himself "the man" means that his revolutionary political spirit carried him through artistic experiments, accidental painting, and the RT instability, which in turn helped inspire Pollock's own revolutionary painting style. Weaving through the story of both artists is the science of fluid dynamics, first as they used it empirically, then through later detailed analysis that illuminates their approaches to painting, their connection to each other, and their effects on art.

Mr. Turner, Artist, Meets
Mrs. Somerville, Scientist

The great 19th-century English artist J. M. W. Turner had an innovative style that earned him the title "the painter of light" and is considered a forerunner of Impressionism. A different aspect is that Turner's works were influenced by the technology of the time and its scientists. One such link was with Mary Fairfax Somerville, a Scotswoman recently honored for her pioneering scientific work by her portrait on a Scottish banknote.

Turner's works vividly portray the sometimes painful transitions due to the technology of the industrial revolution. *The Fighting Temeraire* (1839, National Gallery, London) shows the sailing man-of-war *HMS Temeraire,* looking pale and ghostly as a steam-powered tugboat tows her to be broken up, one of the last ships to have distinguished herself at the Battle of Trafalgar, 1805. In *Rain, Steam and Speed—The Great Western Railway* (1844, National Gallery, London), a train lost in mist but marked by its tall black smokestack hurtles toward the viewer out of an impressionistic swirl of light and color, like technology itself onrushing into an indefinite future.

Turner's work also reflected current science. His biographer James Hamilton comments that at the time, the Royal Society shared a building with the Royal Academy where Turner exhibited, so that artists and scientists readily mingled. Turner was affected by some of the best of his era. Hamilton writes that studies of the sun by the great astronomer William Herschel changed how Turner depicted our central star; that Turner discussed artist's pigments with Michael Faraday, the brilliant researcher in electromagnetism and chemistry; and that Mary Somerville was even inspired to visit Turner's studio [1].

Somerville too represented innovation and a breaking of tradition at a time when few women engaged in scientific careers. Following contemporary practice, her family though comfortably off gave her only a minimal education; but she fell in love with mathematics and astronomy, and studied them intensively on her own. Her first husband left her a widow with an inheritance that let her build a distinguished career while remarrying and raising several children.

Somerville became famous for her skill in expressing scientific ideas. She made Pierre-Simon Laplace's classic astronomical work *Mécanique Céleste*, published in 1829, widely accessible through a double translation, from French into English and from "algebra into common language," as she put it. She also tackled a scientific question that was asked at the time: can light magnetize steel?

To a modern physicist, the answer is "Certainly not!" but the true connection between light and electromagnetism was not known until much later. In 1813, an Italian researcher had reported that he could magnetize steel with violet light. Somerville repeated this experiment with homey materials, bathing a steel sewing needle with violet light from a prism. Her conclusion, published in 1826: "I am induced to believe that the more refrangible rays of the solar spectrum have a magnetic influence." This was later refuted and she showed intellectual honesty by repudiating her own result. Nevertheless, her paper was only the second one by a woman in *Philosophical Transactions of the Royal Society,* and she was later honored in other ways including the new banknote.

Somerville and Turner each left legacies, but did they leave a joint one? The recent biopic *Mr. Turner* (2014) suggests they did, when its writer and director Mike Leigh imaginatively extends Somerville's visit to Turner's studio by having her demonstrate the violet-light experiment. This can be read simply as homage to science and art, but Hamilton proposes a definite influence in Turner's *Ulysses Deriding Polyphemus* (1829, National Gallery, London). In this scene from Homer's *Odyssey,* Turner has painted a violet glow in the northern sky, the direction toward which a magnetic compass needle would point.

We don't know for sure that Turner had Somerville's experiment in mind when he put in that violet sky; but as Hamilton's theory makes clear, what we learn again and again is that science, right or wrong, can inspire artists—in this case, Turner himself, or the movie maker who later interpreted Turner's life, or both.

Reference

1. Hamilton, James. *Turner and the Scientists* (Tate Gallery. London: Tate Gallery Publications, 1998); *Turner* (Random House, New York, 2003).

Connecting with E. M. Forster

As my jetliner rears back, I look up from E. M. Forster's *Howards End* to gaze at the concrete sprawl of airport momentarily filling my window. The rows of parked airplanes and automobiles make a fitting backdrop: In the period when Forster wrote *Howards End*, 1908 to 1910, he was already decrying the filthy, cluttered underside of life in the motorized age. Although he was not alone in despising the stink of gasoline and the frantic pace of vehicles, Forster had an unusual grasp of how technological advance promised to change social interaction—often for the worse.

Forster also had an uncanny ability to predict exactly how technology would develop. At the century's beginning the telephone was new and the computer not even invented, yet Forster anticipated their modern evolution, perhaps most explicitly with his short story "The Machine Stops." Today the Internet and its related technologies are as ubiquitous as the automobile, within easy reach even as I fly five miles up. They raise all sorts of questions about relationships, community, and sexuality—the very same questions that Forster was contemplating in these two works.

For those who have never read *Howards End* (or missed Emma Thompson in the 1992 film version), it is a book about human connection. Margaret Schlegel—the older of the two cultivated, well-to-do sisters central to the story—becomes impassioned over the phrase "Only connect!" which carries two meanings. One is a call to unite the opposing elements within each person—what Margaret calls the beast and the monk, the prose and the passion—while the other is a call to put the greatest energy into personal relations. "Only connect!" is the book's epigraph, and whenever Forster speaks as narrator he emphasizes the value of personal relationships.

But Forster also realizes that the quality of personal connection depends on the quantity—often inversely. "The more people one knows the easier it becomes to replace them," Margaret sighs. "It's one of the curses of London." Too many connections, in other words, devalues each one in a kind of emotional inflation. For the Schlegel sisters, this is the constant danger of frenetic city life; for the characters of "The Machine Stops," it is the inevitable by-product of remote communication technology.

Written in 1909 partly as a rejoinder to H.G. Wells' glorification of science, "The Machine Stops" is set in the far future, when mankind has come to depend on a worldwide Machine for food and housing, communications, and medical care. In return, humanity has abandoned the earth's surface for a life of isolation and immobility. Each person occupies a subterranean hexagonal cell where all bodily needs are met and where faith in the Machine is the chief spiritual prop. People rarely leave their rooms or meet face-to-face; instead they interact through a global web that is part of the Machine. Each cell contains a glowing blue "optic plate" and telephone apparatus, which carry image and sound among individuals and groups.

The story centers around Vashti, who believes in the Machine, and her grown son Kuno, who has serious doubts. Vashti, writes Forster, "knew several thousand people; in certain directions human intercourse had advanced enormously." Although clumsy public gatherings no longer occur, Vashti lectures about her specialty, "Music During the Australian Period," over the web, and her audience responds in the same way. Later she eats, talks to friends, and bathes, all within her room. She finally falls asleep there, but not before she kisses the new Bible, the Book of the Machine.

Kuno, in contrast, once made his way illegally to the surface, where he saw distant hills, grass and ferns, the sun and the night sky. The Machine dragged him back to its buried world, but he understands the difference between pseudo-experience and reality. "I see something like you in this plate," he tells his mother, "but I do not see you. I hear something like you through this telephone, but I do not hear you." Vashti also senses the lack. Gazing at her son's image in the plate, she thinks he looks sad but is unsure ". . .for the Machine did not transmit nuances of expression. It only gave a general idea. . .that was good enough for all practical purposes. . ."

The drama in the story comes as the Machine inexplicably begins to decay, at first producing minor quirks. The symphonies Vashti plays through the Machine develop strange sounds that become worse each time she summons the music. That troubles her, but she comes to accept the noise as part of the composition. A friend's meditations are interrupted by a slight jarring sound, but the friend cannot decide whether that exists in her cell or in her head. More serious problems ensue with food, air, and illumination, but Vashti—

like almost everybody else—clings to the conviction that the Machine will repair itself.

When the Machine's demise is nearly complete, Vashti's faith fails with it in a way that recalls Margaret Schlegel's speech about large numbers of people. For all the thousands Vashti "knows," she dies nearly alone in a mob of panicked strangers, frantically clawing upwards to the Earth's surface as the Machine finally stops. Her sole redemption comes from a moment of true human contact: She encounters Kuno to talk, touch, and kiss "not through the Machine" just before they perish with the rest of the masses. Forster leaves us amid that final failure of the race to "Only connect," sustained solely by the promise that the few who survive on the surface will rebuild, without the Machine.

Long after Forster imagined this dire scene, the technology of connectivity is here. The details differ slightly: Instead of sound and moving images, we exchange written messages and pictures over the Internet, which links computers globally through fiber-optic telephone lines. (I use "Internet" here as a generic term for the major computer webs—the Internet itself and its World Wide Web, and the commercial nets connected to it, such as America Online. These, by the way, can also carry real-time sight and sound, which will surely grow in use.) But the logic of network connectivity, whether one-to-one or one-to-many, remains unchanged, and so does the loss of personal dimensions. The images in Forster's world move and speak, but do not convey facial nuances. Except for the limited use of still images, our electronic messages omit physical attributes. Forster's imaginary system, and our real one, offer unprecedented breadth of connection—there are an estimated 10 to 30 million Internet users worldwide—but do not allow people to touch, or to read each other directly.

On the Internet, Forster's implied questions still beg for answers. Open a magazine or newspaper, and you're likely to find an article asking a simple question: Does the Internet break down isolation, or merely provide pale simulations of friendship and love that drive out the real things? At one extreme, a recent newspaper story announces "On-Line addiction: wire junkies are multiplying as more people withdraw into their private worlds," and describes Internet users who are "ensnared in a net of fiber optic lines . . . and loneliness." But some users tell a different story. Mary Furlong, the founder of a group called SeniorNet, says: "I see a lot of loneliness in the senior population and lack of mobility. . .Going on-line allows you to be intellectually mobile and be socially mobile"—exactly what the optic plates of Forster's world offer its confined inhabitants. There are even indications that the lack of physical presence can be advantageous. A London-based group finds the Internet to be a "nonprejudicial medium," especially valuable for children with conditions like cerebral palsy that affect speech.

In Forster's story, the rise of the Machine came from a belief in progress, which had "come to mean the progress of the Machine." Today, the relentless march of constantly updated computers and infrastructure sweeps us along to use the new technology because it is there, even when it conflicts with existing ways of life. A case in point is occurring in Italy, where a real estate consortium has spent more than $2 million for an entire village, Colletta di Castelbianco, founded in the 13th or 14th century and long since abandoned. The developers will turn its medieval walls and arches into a "telematic" village, creating apartments outfitted with the latest communications equipment including high-speed access to the Internet. The idea is that businessmen can operate on a global scale while enjoying the beauties of the rugged Ligurian region. Even the village cafes will be linked to the Internet, with facilities for video conferencing.

In a nation that values its cappuccino accompanied by enthusiastic conversation, the project evokes mixed reviews. Paolo Ceccarelli, an architect who studies the impact of computers, believes the Italian way of life is unlikely to bring forth many devoted Internet aficionados. He is depressed by the disconnection he sees in the Internet village. Echoing Forster's themes, he contemplates businessmen "parking their BMWs . . . climbing the stairs to their hermetically sealed apartments and plugging in their portables in unison, all blissfully unaware of each other's presence."

It's true that even in Forster's vision, traces of emotion and relationship elude the grip of the Machine. "Human passions still blundered up and down," Forster wrote, and it is clear that Vashti and Kuno share a mother—son link, although Kuno has been raised in a public nursery. Sexual love, however, has changed radically. Sitting passive and isolated, people no longer touch each other, and their physical attractiveness has diminished. Vashti is described as a "swaddled lump of flesh . . . five feet high, with a face as white as a fungus." The Machine controls procreation, sending citizens traveling for the specific purpose of propagating the race. In this world where people do not kiss, where sex happens on assignment, Kuno rails that the Machine "has blurred every human relation and narrowed down love to a carnal act."

Human passions still blunder around the Internet, too, where people have fallen in love and gone on to form complete relationships or marry. But in an outcome that Forster did not consider, the Internet also adds the more or less artificial experiences of cybersex to the sexual repertoire. Sex over the net means incomplete involvement in different degrees, from exchanges among mutually responsive participants to the solitary viewing of pornography. Remote sex has its undeniable impact, and in a world dealing with the disease of AIDS, it may stand in for unsafe actual sex. But can it stand in emotionally for genuine sexual love? Responding to this concern, participants at a recent Vatican conference on "Computers and Feelings" declared that cybersex is "the end of love. It is empty loneliness." That judgment is based on a recognition that human experience is diminished as it is filtered through electronic channels.

Forster went further. Fearing that more technology meant less humanity, he utterly rejected the technical achievements of his time. In 1908, after hearing of the first successful airplane flight over a kilometer-long circuit, he wrote in his journal, ". . . if I live to be old, I shall see the sky as pestilential as the roads. . . Science . . . is enslaving [man] to machines. . .Such a soul as mine will be crushed out." But we have come to learn that instead of producing a monolithically "bad" or "good" effect, a rich technology usually generates a balance sheet of benefits and costs—many of them unpredictable, because people use technology in unexpected ways. Thomas Mann once said, "A great truth is a truth whose opposite is also a great truth." A significant technology also embraces opposites; if some of its applications constrain human potential, others enhance it.

The Internet, in fact, helps people find kindred spirits. Forster did not foresee this development, but his story hints at it, for Vashti could either talk to a friend through the Machine or address an audience. Now many Internet users coalesce into groups that share concerns and emotional affinities. People with unusual beliefs or lifestyles, with secrets they dare not tell family and friends, seek each other out in thousands of news groups, list servers, and chat rooms. These cater to a variety of interests, problems, and ways of life, including a range of strong political views; divorce, grief, and loneliness; and a spectrum of sexuality, from heterosexual to homosexual, lesbian, and bisexual orientations, with variations. Within these groups, the private and the hidden can be revealed and validated, anonymously if desired.

Forster himself grappled with the partial secret of his homosexuality, which colored his life and his writing. His biographer, P. N. Furbank, concludes that Forster knew he was homosexual by the age of 21. But while a gay lifestyle was then acceptable in some quarters, Forster did not feel he could openly declare his sexuality, or act on it freely. The tension remained until he tried to release it in a way that would reaffirm him as a writer. After the success of *Howards End* in 1910, he feared his creativity had dried up. Yet in 1913, the idea for a novel about homosexual love came to him in a moment of revelation. That seemed to show a way out of his barren time, and he wrote *Maurice* enthusiastically and at great speed. When it was done in 1914, however, Forster saw that it could not appear "until my death or England's," and it remained unpublished until after he died.

It is only a speculation, but a revealing one, to imagine how Forster would have fared with access to a like-minded group on a net, where he could have expressed what he had to suppress in the real world. After all, as a student at Cambridge he had been elected to the exclusive intellectual society called the "Apostles," where, among other topics, homosexuality was discussed in a spirit of free and rational inquiry, providing a sense of liberation that Forster later came to value greatly. On-line access would have created the opportunity to circulate *Maurice* to a larger but still select group that would accept its theme—a form of publication that would have brought even greater fulfillment.

Yet time on-line would have been ill used for Forster the writer. Aimless chat is the insidious seduction of the Internet; it can replace inward contemplation and real experience. In the decade after *Maurice*, Forster looked both inward and outward. His internal life became more unified as he came to terms with his self-doubts, and his sexuality. His external life developed as he worked for the Red Cross in Alexandria during the war, returned to England, and left again for his second visit to India. He deepened old relationships, and formed new ones, in all three places. All this must have been necessary, in ways hardly discernible at the time, before Forster could break free of his unproductive period to complete *A Passage to India* in 1924—a ripening that came only through the slow refining of life-as-lived into understanding.

Forster probably would have sensed this—just as he understood technology's potential to both isolate and overwhelm the individual. In the world of the Machine, each person could call or be called through his blue plate; but each could also touch an isolation switch to stop all interchange. While it is not always so simple, we can make individual choices about how and when to use technology. And in allowing his future humans their privacy between bouts of communication, Forster drew a fine metaphor for both aspects of "Only connect!": the joining of beast with monk carried out in the mind's solitude, the essential reaching out to others that breaks isolation—and the combination of the two, through the internal distillation of felt experience.

Science Fiction Covers the Universe and Also Our Own Little Globe

Ever notice how often the alien spaceship lands in Washington, D. C., or New York City rather than Paris, Beijing, or Rio de Janeiro? Since the big science-fiction blockbusters are Hollywood products, it's not surprising that these films are U. S.–centric and it's also true that Washington and New York are major world cities. Even if the aliens want to reach Earthlings via the United Nations that too requires a stop-off in Manhattan.

But CNN headquarters might be a better choice, because there's a whole big globe out there becoming increasingly interconnected by more than the UN International Telecommunication Union, the Internet, and the 24-hour news cycle. International trade and finance, terrorism, global warming, job outsourcing, immigration—all of these are linking people and nations more closely. Like any other cultural product, science fiction must reflect this reality sooner or later.

One early entry in globalized science fiction is an independent film out of Mexico, *Sleep Dealer*, now in limited theatrical release after winning the Alfred P. Sloan Prize at the 2008 Sundance Film Festival. Directed and co-written by Alex Rivera, it explores the impact of technology outside the United States. In an unspecified near future, Memo, a young Mexican man, leaves his home village. The border with the United States has been completely sealed, but that's no problem because technology still allows him to find a job there—sort of. Using metal nodes implanted into his body, Memo "jacks in" to a network from a cyberfactory in Tijuana to remotely operate a machine that's welding a steel skyscraper frame somewhere in San Diego. His fellow cyber workers perform other jobs all over the world.

The political reality underlying this technology is spelled out when Memo's foreman says that the United States "wants our work but doesn't want our workers." When I appeared recently with Alex Rivera on Ira Flatow's NPR show "Science Friday," Rivera related that the film was sparked when he first heard about the outsourcing of U. S. jobs. *Sleep Dealer* extends the idea. In the film, jacking in provides jobs across the border while keeping the first world safely insulated from the third. It also allows a pilot to remotely fly an armed drone to track and kill "aqua-terrorists" suspected of stealing some of the world's dwindling fresh water; and it permits Luz, Memo's love interest, to share her feelings and memories with paid subscribers over the Internet, and with Memo when they have sex.

Technologically speaking, these projections are on track. Jacking in a person to operate devices outside his body by direct neural control has been a reality in the lab for several years. That's now under study for military use by the U. S. Department of Defense. Drone aircraft, though still remotely controlled by a joystick rather than a neural connection, are used daily in Afghanistan and Pakistan by the United States. And consider the astonishing technique of functional magnetic resonance imaging (fMRI), which can correlate signals from the brain with specific thoughts and emotions. Couple this with social networking and the next step will be to download people's deepest feelings to the Web, though maybe not with sex included.

In presenting a possible future as seen by one particular young Mexican, *Sleep Dealer* also presents a wider future that's rushing toward us all. The best literary science fiction has predicted, sometimes decades early, major social trends like the implications of genetic engineering and the impact of the Internet. With some notable exceptions, science-fiction film has been less prophetic and less driven by ideas. *Sleep Dealer* shows that a film that approaches technology from a global perspective can offer a fresh and thoughtful look at a host of today's issues that will continue into tomorrow.

How Realistic Are Movies Set in Space?

One of the oldest themes in science fiction is space travel, where we leave the Earth to explore both the void that surrounds us and other worlds within it. Some critics trace space literature back to classical Greek and other ancient sources. Space travel also has a long history in the modern medium of film, for it was the theme of the first science-fiction film ever made—*Le Voyage dans la Lune (A Trip to the Moon)* by the French film pioneer Georges Méliès. This appeared in 1902, soon after audiences first began watching "moving pictures."

Now, over a century later, after thousands of movies about space travel, people remain entranced by it, as shown by two successful films released in 2013, *Gravity* and *Europa Report*. Taken together, these are uncommonly interesting subjects to study that old question, can feature films deliver real and accurate science, because they represent different approaches to scientific realism in space?

Gravity is a fictional science-based film that is not exactly science fiction, which typically extrapolates known science to provide speculative ideas. But *Gravity*'s story about astronauts dealing with a catastrophic accident takes place within today's space technology—NASA shuttles, spacesuits, and so on. Parts of it could have been shot during a real NASA mission, and the catastrophe itself has real roots.

That contrasts with *Europa Report*, a classic kind of science-fiction story in which a spacecraft carries explorers to Europa, a satellite of Jupiter, to seek life there. We have sent robot spacecraft even further but humans have yet to venture beyond the Moon. Within the speculative possibility of humanity traveling deep into space, however, the film is touted as having a large quotient of real science, and it tells its story in a style that seems authentic.

Comparing these films illuminates how science fares when mixed with entertainment, for there is always tension between solid science and an exciting story. Over the years, space films have handled the balance in different ways.

For instance, before we landed on the Moon in 1969, *Destination Moon* (1950) got space science right. It was based on a Robert Heinlein novel about the first rocket ship to the Moon, and used technical advice from him and German rocket expert Hermann Oberth. The rocket flight is correctly shown and space artist Chesley

Bonestell, known for accuracy, made the Moon's surface look authentic. But as one reviewer put it, the film was "more *sci* than *fi*;" a potentially uplifting story about our first leap into space is instead strangely flat with little drama and unmemorable characters. Still, its earnestness made science fiction more respectable.

Nineteen years later, we reached the Moon and found that real space travel could supply its own excitement, such as motivated the docudrama *Apollo 13* (1995). Like *Gravity,* this film portrays a disaster in space, but a real one. In 1970, NASA's Apollo-13 spacecraft suffered an explosion on the way to the Moon that aborted the mission and caused much anxiety until its three astronauts could be brought safely home. The film dramatized this event while remaining true to space science by realistically presenting the mission and its technology.

NASA took us to the Moon, but speculative science fiction took us further, putting new strains on scientific realism. Still, some films coped. Stanley Kubrick's classic *2001: A Space Odyssey* (1968) imagined a future with routine travel to a space station and the Moon; but when a mysterious black monolith is found and transmits a signal toward Jupiter, the spaceship *Discovery* is sent on a long trip there. *2001* realistically shows microgravity during these trips and also shows that the *Discovery* rotates to provide artificial gravity through centrifugal force, a valid way to combat the physiological effects of microgravity.

Other science fiction like *Star Trek* and *Star Wars* carried us out of the solar system to distant stars and planets. Such tales move the balance between "science" and "story" decidedly toward "story." The reason is that according to known science, we cannot reach the stars in reasonable times. With today's rocket engines, travel to our nearest star other than the Sun would take around 100,000 years. Action plots, however, need short travel times, which implies starships traveling faster than light. But that is impossible within Einstein's relativity, so science fiction invokes imaginary science like *Star Trek*'s "warp drive" to justify fast interstellar trips. This requires suspension of disbelief—a suspension that is willing if the story is good enough.

Scientific realism can also be an issue in films about more modest space travel such as *Gravity,* which opens with two figures in bulky white spacesuits working near a Space Shuttle orbiting only

600 kilometers above the Earth—Matt Kowalski (George Clooney), a veteran astronaut, and Ryan Stone (Sandra Bullock), a medical doctor with limited space experience.

As they change orientations to service the Hubble Space Telescope, the Earth seems overhead, or underneath, or off to one side. Always, it is stunningly big and beautiful; but also an inescapable reminder that humans in space have left their lush natural home for an airless, pitiless void where survival is impossible without technology. That doesn't seem to matter amid the serenity of Earth's beauty and the silence of space, and as Kowalski glides around the Shuttle smoothly propelled by spacesuit thrusters. Like the famous scene in *2001* where a huge space station dances to a Strauss waltz, in *Gravity* a poetic moment softens the starkness of space.

The moment is ruined and the film moves into high gear when Mission Control on Earth abruptly announces that pieces of a destroyed Russian satellite are headed toward the Shuttle. The debris wrecks the Shuttle, kills all aboard, and sends Stone careening off into space. Stunned and helplessly spinning, she panics until Kowalski finds her and tethers her spacesuit to his. Now they're together, but in desperate circumstances: how do they get back to Earth?

Kowalski has an answer: make their way to the International Space Station (ISS), visible some distance away in the same orbit, and use a spacecraft docked there to return to Earth. And so the two set off, with Kowalski's thrusters propelling them as Stone trails along on her tether. When they reach the ISS, Kowalski sacrifices himself so Stone can survive. Alone, she musters the determination and ingenuity to leave the ISS, which has been damaged by the debris storm, reach the Chinese Tiangong space station, and return to Earth in its spacecraft.

This story is similar in spirit to tales of sailors who survive shipwreck at sea and reach land in a lifeboat. Their situation is desperate too but at least their environment is not immediately deadly. The survivors in *Gravity* face the vacuum of space with limited oxygen and depend on high technology to return to Earth.

How does this technology stack up in *Gravity*? Mostly, it is reasonably accurate. Orbital details for the Hubble Telescope are correct, NASA equipment is shown in meticulous detail, and thrusters like Kowalski's were once used. In 1984, NASA tried

Manned Maneuvering Units (MMU) that propelled astronauts by expelling nitrogen gas, but these were deemed unsafe even with a tether (never would NASA let Kowalski fly without one, so chalk up his free movement to artistic license). However, when at one point Stone propels herself with blasts from a fire extinguisher, she demonstrates Newton's Third Law of action–reaction just as well as an MMU.

Other issues are the return to Earth in a "lifeboat" and the story's core, the destruction of the Shuttle by debris. Space lifeboats do exist just as the film shows, in the form of "Soyuz" spacecraft from the Soviet space program of the 1960s that still ferry crew and supplies to the ISS. At least one is always docked there precisely to serve as an escape craft that can make it back to Earth, and a similar craft is moored to the Chinese space station.

As for space debris, it is real. Our planet is surrounded by hundreds of thousands of pieces of space junk, some bigger than 10 centimeters, arising from mishaps with satellites and rockets. These move fast and even a small piece can do damage. Much of the junk was created in 2007 when China tested a missile by destroying one of its own weather satellites, and some of these particular chunks destroyed a small satellite in 2013. So the scenario where a whole Shuttle is destroyed by debris from a satellite blown up by Russia is not wholly imaginary, but it is unlikely; for one thing, pieces from an explosion would probably not move in a compact swarm.

Such chunks might never even come anywhere near the Shuttle. In a big scientific concession, *Gravity* ignores the fact that different man-made objects in space occupy quite different orbits. ISS is not in the same orbit as the Hubble Telescope, but in one some 150 kilometers lower and tilted at a different angle relative to the Earth. Rarely if ever would the two objects be in mutual line of sight and Kowalski and Stone could not possibly make it from one to the other using spacesuit thrusters. This scientific cheat that compresses different orbits into just one generated a degree of outrage from knowledgeable commentators like Neil deGrasse Tyson of the Hayden Planetarium and Dennis Overbye of the New York *Times*. But the story could not work without replacing a crucial piece of real science with Hollywood science.

In contrast to a film set within sight of Earth, *Europa Report* takes us aboard the *Europa One* spacecraft in the "first attempt to send men and women into deep space," a 21-month voyage over hundreds of millions of kilometers to Jupiter's satellite Europa. There the ship's six astronauts and scientists hope to find life in Europa's ocean under a layer of ice.

We follow the mission through onboard cameras that send images back to Earth. *Europa Report* is constructed as if made from this video footage that has been recovered, as revealed at a press conference on Earth held by a Europa mission scientist who helps us piece together a story told in multiple flashbacks.

The recovered video shows life aboard the spacecraft and also an upsetting event, the loss of a crew member as he works in space with engineer Andrei Blok (Michael Nyqvist). But the crew's spirits revive when they land on Europa (though not at the preferred area, the Conamara Chaos, where disrupted ice suggests movement from below) and start drilling through the surface ice to the ocean. Meanwhile, Andrei claims to see a moving light outside the ship but it is unclear if this is real. When the drill finally breaks through, a probe lowered into the ocean transmits images and picks up radiation and thermal activity. Then the image flickers as something hits the probe and all contact is lost.

That leaves one option, a surface walk to the Conamara Chaos—dangerous because of radiation from Jupiter, but marine biologist Katya Petrovna (Karolina Wydra) is eager to go. Her bravery pays off: she finds samples that yield the first alien life, a tiny unicellular organism. With this success, she heads back to the ship but stops to investigate what may be the light that Andrei saw. As she approaches, the light spreads around her, and to the horror of the others her radio signals now come from beneath the ice as she sinks into the ocean, until only static is heard.

The survivors conclude that Petrovna encountered a big life form and decide to return immediately to Earth to announce this, but the ship loses power on liftoff and drops back onto the ice, killing the captain. Andrei and another crew member go outside to repair the engine, but the ice shifts and both disappear into the ocean, surrounded by blue light. With the mission now doomed, the sole survivor, pilot Rosa Dasque (Anamaria Marinca), realizes that all she can do is show the world what has happened. As she sends stored and live video to Earth—the source of the images we have been viewing all along—the ship fills with blue light and as it falls through the ice, we glimpse an immense octopus-like creature in the ocean below.

The bioluminescent alien octopus is pure science fiction but *Europa Report* mostly comes across as realistic. Its "found" video combined with actual newsreel footage gives it the look of a documentary film. The long voyage to Europa believably shows how it would feel to be confined in a cramped and spartan spacecraft (which, as in *2001*, rotates to provide artificial gravity). Most important, the premise that Europa is the likeliest known site for extraterrestrial life is valid, for besides Earth, only Europa seems to contain quantities of liquid water.

Since NASA's 1977 Voyager and 1989 Galileo missions first closely examined Europa, what we continue to learn about it indicates that its icy surface covers a deep ocean that could support life. This interior water is unfrozen, it is thought, because Jupiter's gravity flexes and heats Europa as it orbits, like kneading a ball of clay. The Conamara Chaos is real too, a known feature that would be a natural area to explore. However, the thickness of the surface ice is in question and we do not know if we could really reach the water below as the film implies. Also, the film has one specific scientific omission: it ignores Europa's low gravity, 13% that of Earth.

This lack does not affect the story, whereas in *Gravity*, rearranging the orbits of artificial bodies is critical to the plot. If a film is declared to be scientifically accurate, that raises expectations, and false science in the film can generate a backlash. That is why critics were quick to note the error about orbits as well as other smaller ones, though that has not kept audiences from flocking to *Gravity*. But the act of pointing out errors also presents science to the public. *Gravity* has engendered new awareness of the space program and its issues such as space debris.

Expectations for accuracy are lower for science fiction, which can get away with even far-out speculation. If *Gravity* were a science-fiction story where, say, hostile aliens capture Kowalski and Stone after the Shuttle disaster, followed by an exciting space battle, that would overwhelm any concerns about which space station is in what orbit. *Europa Report*, however, stands out as science fiction that projects mostly plausible speculation. The trade-off though is that some viewers find its low-key style lacking compared to the dazzling special effects in blockbusters like *Star Wars*. Millions enjoy these space spectacles that outdo reality, which is why Hollywood produces so many of them.

But despite their flaws, *Europa Report*, as science fiction, and *Gravity*, as a science-based story, bring satisfying degrees of realism to space while not forgetting to include drama and a human story. More than 60 years after *Destination Moon*, these motion pictures show that it is possible to honor both the science and the fiction.

Hollywood Science: Good for Hollywood, Bad for Science?

Many Hollywood films fit into genres such as action (for example, *The Bourne Legacy*, 2012), crime (*The Girl with the Dragon Tattoo*, 2011), or Western (*True Grit*, 2010), but there is no genre labeled "science" or "technology." Still, these subjects appear in biographical, historical, or documentary films about famous scientists like Madame Curie or momentous events like the development of the atomic bomb. Science has its greatest film presence and impact, however, in the science-fiction and superhero genres. These highly popular motion pictures are big in Hollywood, earning billions of dollars by appealing to huge U. S. and global audiences. As they do, they carry explicit and implicit messages about science and scientists to millions of people.

That raises a question that I first discussed in my book *Hollywood Science* [1] and pursue further here: "Hollywood science" is clearly good for Hollywood, but is it also good for science? Judging by the history and impact of movie science, it can be, if properly used; otherwise, there are risks to science and its credibility.

The beginnings of science-fiction and superhero films

Science fiction has been embedded in film culture almost since movies began, starting in 1902 with director Georges Méliès' *Le Voyage dans la Lune* (*Voyage to the Moon*), from a story by Jules Verne. Some early science-fiction films, such as *Metropolis* (1927) and *Godzilla* (1954), have become classics or at least cult classics; but mostly these motion pictures were low-budget "B" productions with little acclaim until features such as *2001: A Space Odyssey* (1968) and *Star Wars* (1977) brought major commercial and critical success.

Films starring superheroes—imaginary characters whose supernatural or extraordinary abilities are devoted to fighting evil—trace back to the "Golden Age" of comic books, the 1930s to the 1950s. Superman, in many ways the prototype of a superhero, appeared on the comic book scene in 1938, followed by Batman and others. Serialized superhero movies for children came after as did some superhero films in the 1960s; but like science fiction, it took a breakthrough film, *Superman* (1978), to carry the genre to a new level. Now superhero films draw on a whole set of characters such as the Hulk and Spider-Man.

The reach of Hollywood science

Since these beginnings, literally thousands of science-fiction and superhero features have been released [2]. The most successful reach enormous audiences as measured by world-wide box office receipts, and comprise over one-third of the 50 all-time top grossing films; for example *Spider-Man 3* (2007, ranked 23rd with a gross of $891 million), *Star Wars: The Phantom Menace* (1999, 10th at $1,027 million), and *Avatar* (2009, at $2,782 million the all-time highest grossing film ever). Based on ticket prices, each of these films in the top 50 is estimated to have reached from 100 million to well over 300 million viewers, more than the U. S. population [3].

If only because of such huge numbers, these movies are culturally significant. Critics of every cultural and political persuasion routinely analyze what the latest science-fiction or superhero film says about contemporary life and society. *2001* and *Star Wars* appear in lists of all-time best movies. Images, ideas, and language from the two genres have become iconic, from Spider-Man casting a web that lets him nonchalantly swing between skyscrapers, to space travel via *Star Trek*'s "warp drive," to "may the Force be with you" and the Hulk's warning "Don't make me angry! You wouldn't like me when I'm angry!"

Much of the meaning of these films derives from the messages they deliver about science and technology and related societal issues such as climate change.

Science content and accuracy

But how much science really shows up in these films and how valid is it? By definition, a science-fiction film uses scientific ideas, either as currently understood or by speculative extrapolation. For example, *Voyage to the Moon* and *Destination Moon* (1950, based on a story by Robert Heinlein) used existing science to propose that we could reach the Moon. Now that we have, science fiction like *Star Wars* postulates a next stage still far beyond our reach, travel to the stars. Other ideas, conjectures, or events involving every field of science and technology—alien life; earthquakes, disease outbreaks, and other disasters; genetic engineering; artificial intelligence, and more—animate thousands of science-fiction films.

Superhero films do not necessarily involve scientific ideas, but these can appear in superhero origin stories and some superheroes begin as scientists. Their ensuing adventures may also have a "sciency" or hi-tech tinge. Physics major Peter Parker is bitten by a genetically modified spider, becomes Spider-Man, and in *Spider-Man 2* (2004), battles a rogue scientist; Superman can fly because the Earth has a lower gravity than the alien planet of his birth; researcher Bruce Banner is exposed to gamma rays and finds that when he's angry, he turns into the incredibly strong and ferocious Hulk; and industrialist Tony Stark develops a tiny multi-million horsepower "arc reactor" to drive a flying suit that makes him a superhero in *Iron Man, Iron Man 2,* and *The Avengers* (2008, 2010, and 2012).

Whether a given film uses accepted or wildly speculative science, scientists may well have negative views about the science in the film. Some scientists delight in finding the errors, as has been described by astronomer and science blogger Phil Plait. As child and adult, he

> made fun of the science in movies. "That's so fakey!" I would cry out loud when a spaceship roared past...I decided it would be fun to critique the science of movies, and I dove in with both glee and fervor... It was surprisingly easy to deconstruct Hollywood accuracy, or lack thereof. Any mistake was fair game...

(Plait has recently changed his approach) [4]. Filmmakers also note the lack of scientific accuracy. James Cameron, writer and director of *Avatar* and other science-fiction movies, brings an expert's opinion when he says that science-fiction films "almost never get their facts right" [5].

When the science on screen is wrong, scientists tend to think that filmmakers just do not know or care enough to get it right. Yet writers, directors, and actors have their own reasons to avoid error. They want to pull viewers into their imaginary worlds, but blatant violations of scientific ideas or natural law can ruin the illusion. Often enough, however, since film is a visual medium, "getting the science right" comes down only to "getting the science to look right," which means projecting a seemingly authoritative ambience with sleek lab equipment tended by scientists in lab coats.

The biggest factor playing against scientific exactness, though, is that Hollywood is not in business to produce illustrated lectures about chemistry or astrophysics but to turn out entertaining, money making films. That requires narrative drive, a dramatic arc, and compelling characters. When screenwriters or directors apply their story-telling judgment, they may choose to distort or hype the science in order to tell a better story.

The result is continuing tension between story and science that can be well balanced but can also skew a film to either side. As a pioneering movie about spaceflight, *Destination Moon* benefited from having Robert Heinlein consult about the science and artist Chesley Bonestell design the lunar sets, each at the top of his field and known for his commitment to accuracy. The result is scientifically impeccable, with realistic moonscapes and a clever Woody Woodpecker cartoon segment that illustrates the physics of spaceflight. But the story underplays its inherent drama, producing a correct but unexciting film (Fig. 3.1).

Figure 3.1 *Destination Moon* (1950) gets high marks for accuracy in its story of spaceflight to the Moon in an era when that was still science fiction, but lags in drama and excitement.

In other cases, the science may be distorted for story's sake but is not utterly wrong. *The Day After Tomorrow* (2004) presents basic information about global warming such as how scientists trace the history of global temperatures, the effects of warming on sea level and ocean currents, and the counterintuitive fact that the changed currents can cause an ice age. But to generate dramatic heft and a sense of desperate urgency, the time frame is compressed to days and weeks rather than the actual years and decades. The result is an intense, fast-paced story that uses award-winning special effects to both inform and misinform about global warming (Fig. 3.2).

Figure 3.2 While getting some of the science right, *The Day After Tomorrow* (2004) exaggerates and accelerates the effects of global warming, as in this tsunami approaching New York City, rendered by computer-generated imagery (CGI).

Then there are films that deliberately use completely wrong science. The accepted plot device where spacecraft travel faster than light violates the theory of relativity, but viewers are asked to suspend disbelief so that fictional spaceships can quickly cover cosmic distances. Other films stretch nuclear and genetic science beyond reality to produce impossible but striking mutants like the giant dinosaur Godzilla or the human variant superheroes in *X-Men* (2000) and its sequels. Unfortunately, some films contain unforced scientific errors that could just as easily have been expressed correctly without damaging the story, such as misstatements in *The Core* (2003), a film about what happens when the Earth's core stops spinning.

Scientists should certainly point out wrong science when they see it, but they can also fruitfully work out reasonable balances between story and science with filmmakers, or acquaint them with science that might spark movie ideas. The Science and Entertainment Exchange of the National Academy of Sciences (scienceandentertain mentexchange.org) successfully enables such interactions between interested scientists and filmmakers. But as long as science-fiction and superhero films generate billions, there is little reason for Hollywood as a whole to change its approach to science.

If Hollywood science cannot be fully trusted, should we conclude that Hollywood's gain is necessarily science's loss? Not at all, in my opinion. Properly used, fictional science can influence real science by helping to educate students and the public, contributing to the general discourse about science, and even inspiring scientists.

Inspiring careers and stretching imaginations

Soon after NASA's Curiosity rover arrived on Mars in August 2012, the space agency named the rover's landing site "Bradbury Landing" after the late Ray Bradbury, author of the science-fiction classic *The Martian Chronicles*. Asked why Bradbury was so honored, Michael Meyer, the NASA project scientist for Curiosity, replied: "This was not a difficult choice. . .Many of us and millions of other readers were inspired in our lives by stories Ray Bradbury wrote to dream of the possibility of life on Mars" [6].

Fictional science can inspire dreams and encourage scientific imagination. Bradbury first did so though books, and films can do the same. Ask a group of scientists and many will talk fondly about the science-fiction films of their youth. In *The Seven Secrets of How to Think Like a Rocket Scientist,* James Longuski described how space scientists at a "prestigious government laboratory" (probably NASA's Jet Propulsion Laboratory) gathered regularly to watch science-fiction films from the 1950s. They would laugh at the errors, but

they loved these films. They were like children who want to hear the same fairy tale over and over again. These were the fairy tales of the rocket scientists; their unfettered hearts seeking contact with outer space. Their logic turned off . . . their dreams turned on. Imagination wasn't silly to them [7].

Despite the flaws in Hollywood science, what scientists see on the big screen can motivate them to extend what *is* to what *might be*, using real rather than fictional science.

That inspiration can be particularly important for children and adolescents. Their attitudes toward science and scientific careers are influenced by popular culture, especially as it portrays scientists. In one study, Jocelyn Steinke of Western Michigan University and her colleagues pointed out that "most children do not typically come in contact with actual scientists." Instead,

> ...many grow up seeing images of scientists...as depicted by characters and images in books, movies, television programs, magazines, comics, video games, clip art, Web sites, and a variety of other media sources.

The study found that over 40% of 300 students 12 to 13 years old said that their images of scientists came from films or television (the study did not distinguish between the two media) [8].

Unfortunately, these images often convey stereotypes, not fully realized characters. When James Cameron bemoaned inaccurate science-fiction films, he also identified two movie scientist stereotypes as "idiosyncratic nerds or actively the villains." There is also a third less negative stereotype, scientist as hero. Still, a stereotype is a stereotype. All three appear in films and distort the reality of what kinds of people become scientists and why.

For example, mad geneticist Marlon Brando creates unnatural mixtures of human and beast in *The Island of Dr. Moreau* (1996) (Fig. 3.3), and evil geneticist Sean Bean harvests human body parts from clones for profit in *The Island* (2005); climate scientists Dennis Quaid and Ian Holm in *The Day After Tomorrow* heroically risk all (the Ian Holm character dies) to warn the world of coming disaster; Dr. Brackish Okun (Brent Spiner) in *Independence Day* (1996) is unkempt and peculiar; and in an ambivalent, metaphorical turn in *Spider-Man 2*, well-meaning Doctor Octavius becomes destructive Doc Ock (both played by Alfred Molina) through his own science when neural implants he uses to develop clean fusion power take over his mind.

Such stereotypes may give young people only a confused understanding of what it is to be a scientist. However, some movie characters project more balanced images, such as radio astronomer

Ellie Arroway in *Contact* (1997) (Fig. 3.4). Jodie Foster plays the part well and the film convincingly shows Arroway's early interest in science as fostered by her father, the rewards of a commitment to science along with the difficulty of balancing it with a personal life, and Arroway's scientific integrity—all parts of being a real scientist. Anecdotal evidence and analyses show that scientists, non-scientists, and media scholars alike find Ellie Arroway an appealing and valid scientist role model [9–11].

Figure 3.3 Geneticist Dr. Moreau (Marlon Brando) in *The Island of Dr. Moreau* (1996, the latest film version of the H. G. Wells story), is a mad scientist who creates a race of half-human monsters.

As a bonus, this realistic character is female, not male. The historically low number of women in science and engineering is growing but they remain underrepresented [12]. As Steinke notes, one approach to increasing their numbers is to give females opportunities to encounter women scientists:

> Interaction with women scientist role models has been singled out as an important factor in fostering positive attitudes toward science and scientific careers in girls and young women...In the absence of real-life role models, images of women scientists in the media may serve as important sources of information...[13]

Figure 3.4 Radio astronomer Ellie Arroway (Jodie Foster) in *Contact* (1997) is a relatively realistic and nuanced depiction of a scientist on screen.

Films can provide this inspiration since women scientists are increasingly appearing on screen. For instance, in 60 movies made from 1929 to 2003, only 18% of the scientists portrayed were female [14, 15]. Similarly, in the Internet Movie Data Base IMDB. com I found that in over 800 films with scientist characters released from 1915 to 2012, fewer than 20% included women scientists. But more recent films released from 1991 to 2001 show a big increase to 34% females among scientist and engineer characters [16]. This correlates with growth from 1993 to 2008 in the female fraction of U. S. workers with science or engineering degrees (21% to 28%) or with careers in those areas (31% to 38%) [12]. It is unclear, however, how much of this was influenced by movie characters, especially since few scientists are portrayed as well as in *Contact*.

Beyond potentially interesting young people in science careers, Hollywood science can help them learn real science—remarkably, even if the Hollywood science is wrong.

Educating scientists and non-scientists

Essayist Susan Sontag once noted that science-fiction films wield a power of "sensuous elaboration. . .by means of images and sounds. . ." that gives them an impact far beyond written science fiction" [17]. Today, digitized scenes created by computer—dubbed CGI, computer-generated imagery—provide visual power that can give these films a big role in teaching science.

First used in the robot story *Westworld* (1973), CGI can show every kind of advanced technology, distant planet, and alien being that inventive writers can create, and can simulate exotic phenomena like tsunamis and black holes (see Fig. 3.2). In our media-driven age with students attuned to visual representation, these effects can enhance science teaching, especially if combined with absorbing emotion and drama (which may be a problem. Some film critics comment that unimaginative reliance on CGI coupled with a certain corporate style of film making is producing formulaic movies that lack feeling, meaning, and characterization, including many superhero films) [18, 19].

Whatever the cinematic merits, one teaching approach is to enrich regular science courses with lessons from the movies, as Costas Efthimiou at the University of Central Florida has pioneered in "Physics in Films." For example, he uses *Armageddon* (1998), in which a huge asteroid "the size of Texas" is on a collision course with the Earth. To avert this, NASA sends up a team of oil drillers to plant a thermonuclear bomb deep inside the oncoming rock. When the bomb is set off, it will presumably split the asteroid and push the two pieces sideways to bypass the Earth on either side.

In the movie, this plan saves our planet, but does real science support the happy Hollywood ending? To find out, the students calculate the asteroid's mass, use the laws of mechanics to compute the paths of the two halves, and reach a surprising conclusion: even a powerful hydrogen bomb explosion would separate the two massive chunks of rock by only 400 meters. Both would still hit the Earth, and disaster would not be averted after all. This example engages the students and then channels their interest to exercise their analytical and scientific abilities. The value of the approach is borne out by the fact that students in the course perform better than those in a similar course without films [20, 21].

A course I have co-taught since 2006 at Emory with my colleague Eddie von Mueller of Film Studies exemplifies a different approach [22]. Rather than work within an existing science course, "Science in Film" was designed to illuminate both science and cinema for science and humanities students, as expressed in the two course texts: my own *Hollywood Science*, about the science in movies, and Vivian Sobchack's *Screening Space*, about the cinematic and cultural meanings of science fiction [23].

The course is built around films chosen to cover important scientific ideas and events and their human impact, from climate change to the rise of computers, or that show scientists. Students hear lectures about the science, given in broad terms rather than quantitative detail, and watch the associated films. Having absorbed all this, the students are led by both instructors to discuss the real and the fictional science and are also required to write film logs and papers.

The course enhances science literacy by introducing big scientific ideas and also through its cinematic perspective. The students are asked to consider how a film packages its science and scientists— what choices are made, what attitudes and agendas underlie them. This encourages critical thought about varied approaches to science in outlets from entertainment and news media to research journals. That makes students better able to thoughtfully weigh scientific claims, for instance in public policy debates, where film itself can also play a role.

Public discourse

Some science-fiction or superhero films influence public discussion of science because they express a particular viewpoint. Others do so as products of their times that reflect existing values and concerns. As they echo and amplify current issues, their wide exposure makes them one more voice that discusses science in our lives.

These influences were apparent in the 1950s and 1960s during the Cold War when the world faced serious nuclear threat. The classics *The Thing from Another World* (1951) and *The Day the Earth Stood Still* (1951) commented on nuclear dangers, *Godzilla* conveyed fears of radiation, and *On the Beach* (1959) and *Dr. Strangelove* (1964) dealt with nuclear war. Later *The China Syndrome* (1979) presented a fictional civilian nuclear accident nearly simultaneously with a real reactor meltdown at Three Mile Island, and *The Sum of All Fears* (2002) was about nuclear terrorism. Now, nearly 70 years after the first nuclear explosion, films still use radiation or nuclear angst as a *deus ex machina* and in their back stories, as in the origins of the Hulk, and of superhero Dr. Manhattan in *Watchmen* (2009).

Many films express new concerns about genetic engineering and cloning. After the structure of DNA was determined in 1953, genetic manipulation started to become a reality. It entered into *The Boys From Brazil* (1978), *Jurassic Park* (1993), *The Island of Dr. Moreau* (1996), *The Sixth Day* (2000), *The Island* (2005), *Splice* (2009), and more. Most of these films are pessimistic about evil geneticists or societies based on genetic discrimination—the theme of *Gattaca* (1997)—and some tackle the morality of genetic engineering.

Tension between science and religion also enters into a recent high-profile film, *Prometheus* (2012). This semi-prequel to *Alien* (1997) has flaws along with high points, but with its story about the origins of life, a scientist character with religious convictions, and a near-human android character, it evokes discussion about how and why life was created (Fig. 3.5).

Figure 3.5 By speculating about how life began on Earth and including a scientist with religious faith (Noomi Rapace, center) and an artificial human (Michael Fassbinder, right), *Prometheus* (2012) invites thoughts about how, why, and by whom humanity was created.

Among films that have strongly affected public consciousness, the dramatically enhanced climate change presented in *The Day*

After Tomorrow created much controversy about the film's scientific accuracy and political influence. After it opened in 2004, Anthony Leiserowitz of the University of Oregon (now Director, Yale Project on Climate Change Communication), surveyed 529 adults and found that the film had "significant impact." It led viewers to

> higher levels of concern and worry about global warming [and] encouraged watchers to engage in personal, political, and social action to address climate change and to elevate global warming as a national priority . . . the movie even appears to have influenced voter preferences.

This influence was widespread. Domestic (United States and Canada) filmgoers bought 30 million tickets, and worldwide box office sales of $544 million represent many more millions of viewers [24]. Also, during the weeks before and after its release, internet traffic to websites about global warming grew significantly, probably aided by the major marketing effort for the film [25].

It is illuminating to compare this film to a less sensational documentary released two years later. In *An Inconvenient Truth* (2006), former Vice-President Al Gore made the case for global warming using mostly accepted science without major distortion or dazzling CGI, essentially as an illustrated lecture. The film won a Best Documentary Academy Award and grossed $25 million each domestically and overseas—excellent for a documentary, but representing fewer than 4 million domestic tickets compared to 30 million for *The Day After Tomorrow*.

This disparity shows in no uncertain terms that distorted science reaches more people than real science, which returns us to the original question "Is Hollywood science good for science?" The question needs an answer because science-fiction and superhero films will continue to reach big audiences, with all that implies.

The future of Hollywood science

Hollywood sees a bright future in fictional science. IMDB.com lists about 200 new science-fiction and superhero titles, mostly Hollywood products, due for release in 2013–2016 or in development. Some will likely be major hits, such as continuations of successful franchises like *Iron Man 3, RoboCop, The Hunger Games: Catching Fire*, and *Star Trek Into Darkness* in 2013; *Jurassic Park IV* and *X-Men: Days of Future Past* in 2014, and *The Avengers 2* in 2015.

These upcoming films carry assorted possibilities for real science. Most are science fiction rather than superhero stories, where science is often less prominent or may be displayed along with magical or divine powers. In *The Avengers*, for instance, the supernatural abilities of the gods Thor and Loki are confusingly featured on an equal basis with Iron Man's flying suit and other hi-tech elements (Fig. 3.6). But this scenario too can be used to teach science, or of equal value, the distinction between science and magic. Some of the films also cover big issues like climate control (also known as geoengineering), a controversial method under study by scientists and policy makers to reverse climate change, and so could produce an impact like that of *The Day After Tomorrow*.

Figure 3.6 Some superhero films mix scientific or hi-tech elements with fantastic or supernatural ones. *The Avengers* (2012) features both Iron Man in his advanced flying suit (right, Robert Downey Jr.) and the god Thor with his magic hammer (second from left, Chris Hemsworth), along with other super characters.

The Day After Tomorrow is eight years old and *Contact* came out in 1997. It is striking that many of the films that convey good science or plausible scientists, or inspire discussion, date back years. These older movies would occasionally pause the action to deliver some exposition, exercising viewer's minds even if the science was fanciful. CGI now tempts filmmakers to replace exposition with non-stop spectacle. If fantasy hugely (sometimes numbingly) amplified by computerized filmmaking is "chasing human temperament and destiny—what we used to call drama—from the movies" as David Denby writes, it is also chasing away ideas [26].

Still, it may take spectacles to attract millions of viewers and give a scientific issue wide circulation. Scientists have to face the quandary that though high popularity for a film translates into broad exposure for its science, it probably correlates with low accuracy. The dilemma is well put by Ron Von Burg at Christopher Newport University, who studies how science is communicated [27]:

> [H]ow can scientists marshal the increased public attention that accompanies a popular film to help communicate important scientific matters to non-scientists without undermining their scientific credibility? [28]

There is no single or simple answer, but scientists can use film to their advantage if they remember that movies are not lectures in Science 101. If scientists look for the true scientific nugget behind even outlandish screen science, and express it accurately within the sense of wonder that Hollywood can create, the result can be inspirational power, better teaching, and greater outreach.

Like any moviegoer with popcorn in hand, any scientist can enjoy watching a science-fiction classic from the 1950s or Hollywood's latest superhero effort; then he or she can step into a classroom or behind a podium and draw on what the film offers in drama or visual richness or excitement to make meaningful points about, and for, science.

References

1. Perkowitz, S. *Hollywood Science: Movies, Science and the End of the World*; Columbia University Press: New York, 2010.
2. The Internet Movie Data Base IMDB.com lists over 6,000 feature films in the genre "sci-fi" or under the keyword "superhero" ("superhero" is not listed as a stand-alone genre), with some overlap between the two areas and with the related genres of "fantasy" and "horror."

3. Box office and ticket price data from http://boxofficemojo.com. Box office figures not adjusted for inflation.

4. Plait, P. How I Stopped Worrying About Science Accuracy and Learned to Love the Story. *The Science and Entertainment Exchange,* May 24, 2012.

5. Revkin, A. Filmmaker Employs the Arts to Promote the Sciences. *The New York Times,* Feb. 1, 2005, p. D2.

6. Vejvoda, J. NASA Names Mars Curiosity Landing Site After Ray Bradbury. *IGN,* Aug. 23, 2012.

7. Longuski, J. *The Seven Secrets of How to Think Like a Rocket Scientist*; Springer: New York, 2010; pp. 5–6.

8. Steinke, J., et al. Assessing Media Influences on Middle School-Aged Children's Perceptions of Women in Science Using the Draw-A-Scientist Test. *Science Communication* **29**, 1, Sept. 2007, pp. 35–64.

9. Steinke, J. Women Scientist Role Models on Screen: A Case Study of *Contact. Science Communication* **21**, 2 (1999), pp. 111–136.

10. Perkowitz, S. Hollywood Physics. *Physics World*, July 2006, pp. 18–23.

11. Reference 1, pp. 181–184.

12. National Science Board. *Science and Engineering Indicators 2012*; Arlington VA: National Science Foundation (NSB 12-01).

13. Reference 9, pp. 111–112.

14. Flicker, E. Between Brains and Breasts-Women Scientists in Fiction Film. *Public Understanding of Science* **12**, 3, July 2003, pp. 307–318.

15. Flicker, E. Representation of Women Scientists in Feature Films: 1929 to 2003. *Bridges*, Vol. 5, April 14, 2005.

16. Steinke, J. Cultural Representations of Gender and Science: Portrayals of Female Scientists and Engineers in Popular Films. *Science Communication* **27**, 1, Sept. 2005, pp. 27–63.

17. Sontag, S. *Against Interpretation and Other Essays*; Picador: New York, 2001; p. 212.

18. Dargis, M. and A. O. Scott. Super-Dreams of an Alternate World Order. *The New York Times.* June 27, 2012.

19. Denby, D. Has Hollywood Murdered the Movies? *The New Republic*, Oct. 4, 2012, pp. 29–40.

20. Efthimiou, C. and Llewellyn, R. Cinema as a Tool for Science Literacy, *Arxiv*, 16 April 2004.

21. Efthimiou, C., and Llewellyn, R. Avatars of Hollywood in Physical Science. *The Physics Teacher* **2006,** 44, pp. 28–33.

22. Perkowitz, S.; von Mueller, E. Communicating Real Science Through Hollywood Science. In *Taking Science to the People*, C. Johnson, C., Editor; University of Nebraska Press, Lincoln, 2010; pp. 81–88.

23. Sobchack, V. *Screening Space: The American Science Fiction Film*; Rutgers University Press, New Brunswick, NJ, 1997.

24. Leiserowitz, A. Before and After *The Day After Tomorrow*. *Environment* **46**, 9, Nov. 2004, pp. 22–37.

25. Hart, P. and A. Leiserowitz. Finding the Teachable Moment: An Analysis of Information-Seeking Behavior on Global Warming Related Websites during the Release of *The Day After Tomorrow*. *Environmental Communication* **3**, 3 (2009), pp. 355–366.

26. Reference 19, p. 33.

27. Von Burg, R. Decades Away or The Day After Tomorrow? Rhetoric, Global Warming and Science Fiction Film. *Critical Studies in Media Communication* **29**, 1 (2012), pp. 7–26.

28. Von Burg, R. Decades Away or The Day After Tomorrow? *Communication Currents* **7**, Feb. 1, 2012.

Turing and Hawking, Typical Nerds?

Mad scientists have a long history in film and TV, most famously in *Frankenstein* (1931), but other strange scientists have appeared on screen before that, such as the wizard-like character Rotwang in the classic film *Metropolis* (1927), who builds a female robot that displays an unsettling sexuality; or Dr. Jekyll, who concocts a potion that turns him into the bestial Mr. Hyde, with results portrayed in various *Dr. Jekyll and Mr. Hyde* films from 1908 to 1920. Today however scientists are typically shown, not as truly mad or destructive, but as eccentric or "different." That describes Sheldon Cooper in the TV sitcom *The Big Bang Theory,* a brilliant theoretical physicist whose cluelessness about human relations makes him funny and endearing.

But does this fictional character capture something meaningful about scientists? Or does Sheldon merely embody a popular stereotype that even colors how real scientists are portrayed? These questions arise in two recent popular and acclaimed films, *The Imitation Game* and *The Theory of Everything,* that between them garnered thirteen Academy Award nominations in 2014 including Best Picture for each. These films portray two important, well-known real scientists, Alan Turing and Stephen Hawking respectively, whose achievements along with compelling life stories make them good choices for semi-biographical films. Turing, a gay man at a time when homosexual activity was criminalized in his native England, died an apparent suicide in 1954; Hawking has for most of his adult life suffered from a serious disabling disease.

Turing, the lead character in *The Imitation Game,* is the mathematician who helped break the German Enigma code in World War II. This dauntingly complex code was used by the German military for its war-time communications. Decoding it required a long-term effort that began with Polish cryptographers in 1932 and was later carried out by English codebreakers at Bletchley Park, an estate northwest of London. Historians estimate that the intelligence gained by secretly deciphering enemy messages in what was called the Ultra project shortened the war by years and may even have provided the edge that defeated Nazi Germany.

Besides his war work at Bletchley, Turing was a founder of modern computational science and the inventor of the so-called Turing test. That determines if a machine is displaying real intelligence by checking whether it can successfully imitate a human in conversation with a person carried out via written messages. This goal has yet to be reached, but a kind of simplified Turing test shows up when you log on to a website where you must answer a question to confirm that you are a person, not a computer. The input question, which might typically ask you to interpret a string of numbers displayed pictorially, is called a "captcha," that is, a "completely automated public Turing test to tell computers and humans apart."

Turing's original name for the conversation test was the Imitation Game, but the title of the film does not refer to Turing's work in computational science, which the film barely mentions. Rather, it is a hint about the unconventional and hidden aspects of Turing's life as a gay man in a society that did not accept open homosexuality.

Stephen Hawking in *The Theory of Everything* also has a life story that is far from ordinary. He is the British mathematician and theoretical physicist who wrote *A Brief History of Time*, which has sold over ten million copies, and has done ground-breaking fundamental research in cosmology and the origins of the universe, quantum mechanics, and general relativity. He is especially interested in black holes, those regions of spacetime where gravity is so intense that neither light nor matter can escape; or at least that was the understanding until Hawking showed that there may be a quantum escape route for energy and information to leave a black hole. These topics are all part of physicists' search for a Theory of Everything that can explain the entire Universe from its smallest to its largest scales; hence the title of the film.

In his twenties, Hawking was diagnosed with the motor-neuron disease amyotrophic lateral sclerosis (ALS, also called Lou Gehrig's disease after the famous baseball player who contracted it). This had almost completely physically disabled him, but he outlived medical predictions of his early death until he passed away in 2018, and his condition did not dampen his scientific creativity and achievement.

The dramatic elements of Turing's and Hawking's lives would seem enough to motivate gripping films about them just as they are. But the reality is that whether a film is about actual scientists like these two, or the imaginary scientists who appear in science-fiction stories, scientist stereotypes tend to endure. Few people, including the writers, directors, and actors who bring scientists to the screen, have met any real researchers, who are only a tiny sliver of the general population—at a rough estimate in the United States, only about one out of every 300 people.

That unfamiliarity makes it easy to pigeonhole scientists as socially awkward or disengaged, though that image may contain a nugget of truth. Some studies by Simon Baron-Cohen, a psychologist at Cambridge University, suggest correlations between scientific ability and tendencies like a limited sense of empathy found in Asperger's syndrome, at the mild end of the autism spectrum. But apart from any claims that scientists suffer from Asperger's, anyone who knows scientists knows they take their work extremely seriously. To researchers eager to probe the cosmos or cure cancer, science is not just a job, it's an all-absorbing calling. By comparison, social conventions such as the rituals that lubricate human interactions can seem unimportant. Such attitudes can easily translate into social or emotional detachment.

Of course, scientists manage to have friends, relationships, and emotional lives; but their characteristic intensity suggests the origins of a fictional scientist like Sheldon Cooper, who disdains chitchat, can be painfully frank, and is impatient with lesser minds. In *The Imitation Game,* Alan Turing (played by Benedict Cumberbatch) displays similar traits, beginning with his crucial interview to join the Enigma project. He is oblivious to the niceties of a job interview, shows little deference to his interviewer, and completely misses a simple joke the interviewer makes. Turing seemingly operates at a serious and literal level that excludes any kind of humor. To make him even less someone you would ever want to have a beer with, he is also arrogantly certain that he can single-handedly break the code and openly contemptuous of the mathematicians and linguists who are his coworkers at Bletchley.

But as should happen in every good story, Turing undergoes some character development after encountering a twin dose of reality: the code breaking computer he builds does not work well, and he meets Joan Clarke (Keira Knightley). She's a pretty young woman, but more important, she is also the smartest person among many applicants to join the code breakers. After Turing adds her to his team, they become friends and even contemplate a platonic marriage.

Fortunately, Clarke has her share of human empathy along with mathematical ability. She understands that Turing's need for secrecy and his "imitation game" take their toll, and that he is lonely. She also sees how the barriers he has put up keep him from working with his group to solve Enigma. To help, Clarke provides him with some simple, straightforward advice about connecting with people; give them a present and tell them a joke. Turing hands out apples to his group and earnestly relates an old gag about running from a bear in the woods. It hardly gets a laugh from the others, but that is when the ice melts, the team meshes, and with Clarke's help, works together to make Turing's machine finally break the code.

Though Turing's joke failed in the movie, the scene worked well, judging by the reaction at the showing I attended. The audience loved watching the robotic, obnoxious Turing try to humanize himself, much as Sheldon's struggles with human contact get huge laughs on *The Big Bang Theory.* This scene was good movie making in that it engaged the audience; but the question is, was it true to Turing?

In an essay in *The New York Review of Books*, Christian Caryl thinks not. The film's director Morten Tyldum and screenwriter Graham Moore, Caryl writes, have reduced Turing to "the familiar stereotype of the otherworldly nerd" and "the caricature of a tortured genius." According to Turing's biographers, Caryl goes on, the mathematician was eccentric, but not utterly remote and unlovable as in the film. He could be "a wonderfully engaging character. . .notably popular with children and thoroughly charming to anyone for whom he developed a fondness," and he had a "sprightly sense of humor."

If this is the real Turing, why did the filmmakers portray him otherwise? In Caryl's reading, they wanted to heighten the tension between Turing and society that arises because Turing is a "martyr of a homophobic establishment." Besides showing only unattractive parts of Turing's personality, Caryl points to other features of the film that make Turing a victim, such as how it presents his eventual suicide. This is described as if due to his misery under chemical castration, the legal punishment decreed when Turing's homosexual activities are discovered. But other evidence suggests that this punishment did not at all send him into suicidal depression. Indeed, there is some evidence that Turing's death may have been a misdiagnosed fatal accident, though the true facts about his death may never be known.

Our knowledge of Turing, correct or incorrect, comes from recollection and biography, most notably what is considered the definitive treatment, *Alan Turing: The Enigma* by Andrew Hodges, the basis of *The Imitation Game.* If Turing were to object to his resulting depiction in the film, he is no longer alive to say so. Unfortunately, Stephen Hawking is also no longer with us, but *The Theory of Everything* is based on a book by someone who knew him intimately, his ex-wife Jane Hawking (she has kept her married name). Her book *Traveling to Infinity: My Life with Stephen Hawking* is the starting point for the film and our view of Hawking as interpreted by screenwriter Anthony McCarten, director James March, and actor Eddie Redmayne, who plays the scientist.

Like the Turing film, *The Theory of Everything* is about a scientist who cannot live exactly according to society's expectations, not because of sexual orientation, but because of his physical difficulties. If *The Imitation Game* makes Alan Turing too much the victim, *The Theory of Everything* shows Stephen Hawking as strenuously trying to avoid that label.

As the film opens, we see Stephen as an active young man agilely riding a bicycle through the streets of Cambridge in the early 1960s, where he is a graduate student at the university, on the way to a party—which turns out to a most important party, because it is where he meets his wife to be, arts student Jane Wilde (Felicity Jones). Later, as Turing takes Jane to the formal university May ball, he dazzles her with well-chosen scientific facts, such as why the men's white dress shirts glow under the ball's black light illumination. By the time we see the two outlined against a stunning fireworks display, as if looking out together into the starry universe that Stephen's research explores, we know a romance is underway. They soon become engaged, and marry in 1965, shortly after Stephen receives the devastating diagnosis of ALS.

As their married life progresses, Stephen's scientific achievements and fame grow while his physical abilities decline, until he reached his final condition, paralyzed, confined to a wheelchair and able to speak only through a voice synthesizer. But despite difficult and terrifying moments as the disease develops, he generally does not see himself as a victim and fights to maintain as much independence as he can. Despite this admirable attitude, however, Stephen requires constant assistance from Jane, their friends, and attending nurses.

It is only this support that enables Stephen to survive and function, and beyond that, to devote himself fully to physics; but though Jane tries hard, she finds it increasingly difficult to care for him and their three children. Tension develops between the two, Stephen leaves her in 1990, and they divorce in 1995. Both remarry— Stephen to one of his nurses (though they too later divorce), Jane to a family friend—yet the film ends with a reconciliation of sorts; when Stephen is honored by Queen Elizabeth as a Companion of Honour, it is Jane who he asks to accompany him to the awards ceremony.

Is this story an accurate depiction of Stephen, whose life and scientific success were so dependent on Jane? Unlike Turing and *The Imitation Game*, these principals can speak for themselves. When he was alive, Stephen had endorsed the film as "broadly true," and Jane has called it "beautiful," adding that Felicity Jones played her well: ". . .she had my mannerisms, she had my speech patterns." We the movie audience also see a seemingly realistic mix of elements in their lives: moments when anger at his fate gets the best of Stephen, and when Jane's frustration at the difficulties she faces get the best of her; other moments when Stephen's wit and intelligence charm Jane; yet others when Stephen savors the science that engrosses him and the acclaim that it brings, while Jane must balance pride in his success with the knowledge that it could not have happened without her, and the desire to earn her own recognition.

This is not to say that the film is utterly realistic. After calling *The Theory of Everything* "beautiful," Jane added "I had to reconcile myself to the compromises that one has to make for the film industry." As is true of *The Imitation Game* as well and of all biopics, the filmmakers faced the difficulties of compressing a whole life into a mere two hours, and felt the need to alter the real story and real characters for dramatic effect—which, after all, is what sells movie tickets.

Though it is refreshing to see attempts to depict scientists on screen as real people, both films have their distortions. For instance, the final scene in *The Theory of Everything* where Jane accompanies Stephen to be honored by the Queen really occurred while they were still married. The film has also inspired negative comment about its treatment of disability and for diminishing Jane and her role. But the image of Stephen himself, a man with great intellectual gifts whose body betrays him, rings true. In living so much inside his head, he could be called the ultimate scientific nerd; but he is no stereotype, because we are given something of a rounded portrait and insight into the character of this particular and unique person.

The Imitation Game takes much greater liberties with the truth, including a wholly fictional subplot that has Turing fail to reveal the identity of a Soviet spy to the authorities, in order to protect his own secret. Turning Turing into a traitor because of his homosexuality is one more problem with the film's depiction of his sexual orientation.

My main point though is that even had Turing been heterosexual, to the moviemakers he would still apparently have been that oddity, a scientist. As I wrote at the beginning of this piece, scientists on screen are typically packaged in stereotypical form, which is easy to do. Unfortunately, the makers of *The Imitation Game* did not resist that temptation. Instead of the real, complex, and ferociously brilliant Alan Turing, they gave us just one more movie scientist, an otherworldly nerd not all that different from Sheldon Cooper in *The Big Bang Theory*.

Boldly Going for 50 Years

Half a century ago, in September 1966, the first episode of *Star Trek* aired on the U. S. television network NBC. NASA was still three years short of landing people on the Moon, yet the innovative series was soon zipping viewers light years beyond the Solar System every week. After a few hiccups it gained cult status, along with the inimitable crew of the starship USS *Enterprise*, led by Captain James T. Kirk (William Shatner). It went into syndication and spawned 6 television series up to 2005; there are now also 13 feature films, with *Star Trek Beyond* debuting in 2016.

Part of Star Trek's enduring magic is its winning mix of 23rd-century technology and the recognizable diversity and complexity enshrined in the beings—human and otherwise—created by the show's originator Gene Roddenberry and his writers. As Roddenberry put it, "We stress humanity." The series wore its ethics on its sleeve at a time when the Vietnam War was raging and anti-war protests were proliferating, along with racial tensions that culminated in major U. S. urban riots in 1967–1968. Roddenberry's United Federation of Planets, a kind of galactic United Nations, is an advanced society wielding advanced technology, and the non-militaristic aims of the *Enterprise* are intoned at the beginning of every episode in the original series (TOS): "To explore strange new worlds; to seek out new life and new civilizations; to boldly go where no man [later, 'no one'] has gone before."

Over the decades, Star Trek technologies have fired the imaginations of physicists, engineers, and roboticists. Perhaps the most intriguing innovation is the warp drive, the propulsion system that surrounds the *Enterprise* with a bubble of distorted spacetime and moves the craft faster than light to traverse light years in days or weeks. In 1994, theoretical physicist Miguel Alcubierre showed that such a bubble is possible within Albert Einstein's general theory of relativity, but would demand massive amounts of negative energy, also known as exotic matter [1]. This is not known to exist except (possibly) in minuscule quantities; and some physicists speculate that the Alcubierre drive might annihilate the destined star system. The warp drive remains imaginary—for now.

However, another application of warped spacetime in the series has been realized: a cloaking device that shields spacecraft from view by bending light around them. In 2006, electrical engineers David Smith and David Schurig built a "metamaterial" electromagnetic cloak that hid an object from microwaves by refracting them to pass around it, much as water flows around an obstacle [2]. Now, similar diversionary tactics are being used to hide small objects under visible light, for instance by electrical engineer Xingjie Ni and his colleagues, who devised a "skin cloak" 80 nanometers thick to do the job [3].

The exotic *Enterprise* transporter, which instantaneously dematerializes and teleports people and things (inspiring the catchphrase "Beam me up"), was supposedly conceived to save the costs of staging repeated spaceship landings. It has a real analogue in quantum teleportation. In 2015, for instance, quantum optics researcher Hiroki Takesue and his colleagues harnessed entanglement to send the properties of one photon to another over 100 kilometers of optical fiber [4]. Above the atomic level, however, we're a long way from teleporting entire organisms or objects.

Other Star Trek technologies anticipated modern trends. The tricorder that TOS medic Leonard 'Bones' McCoy (DeForest Kelley) uses for diagnosis has spawned real devices, such as SCOUT from medical-technology company Scanadu in Moffett Field, California. Meanwhile, activity trackers already perform basic health monitoring, recording pulse rate, calorie intake, and quality of sleep.

Artificial intelligence has begun to emerge in technologies such as speech recognition by Apple's personal-assistant program Siri, Google's self-driving car, and the "all-terrain" Atlas robot created for the U. S. Defense Advanced Research Projects Agency. All are significant developments that could pave the way to an eventual approximation of Lieutenant Commander Data (Brent Spiner), the sentient android who debuted on television series *The Next Generation* in the late 1980s.

Star Trek's holodeck—the immersive virtual-reality environment in which the *Enterprise* crew visits simulated locales—is also years away, but huge advances in the technology are afoot. The Oculus Rift headset, for instance, provides a visual and auditory virtual-reality experience, but must be tethered to a computer, thus falling short of delivering the seamless holodeck experience.

Three-dimensional printers, which lay down successive layers of material to form intricate shapes, are now being adapted to handle food, perhaps a step towards *Enterprise* meal replicators. The Creative Machines Lab, then at Cornell University in Ithaca, New York, designed one model as part of its open-access Fab@Home project, and Natural Machines in Barcelona, Spain, touts its Foodini printer as simplifying the making of textured or layered foods such as ravioli.

More generally, and arguably with greater long-term significance, Star Trek raised enthusiasm for space exploration and science. In 1975, fans convinced NASA to name its first test space shuttle orbiter *Enterprise* (the craft was unpowered and never reached space). And many young would-be scientists have found the series inspirational.

Its social message has been no less important. The federation ethic ensured that Kirk, *Next Generation* Captain Jean-Luc Picard (Patrick Stewart), and their successors "waged peace" even when confronted by aliens such as the Klingons, a people genetically predisposed to hostility. The February 1968 episode "A Private Little War," an allegory about Vietnam, was a pointed example. Roddenberry believed that humanity must learn to delight in difference, even in alien life-forms, and ready itself to "meet the diversity that is almost certainly out there."

Star Trek's portrayal of human diversity and refusal to engage in national exceptionalism remain landmark achievements. Emerging at a time of racial exclusion in U. S. television, TOS crew included Lieutenant Nyota Uhura (Nichelle Nichols), the first prominent African American female role in a U. S. television series, as well as the "pan-Asian" helmsman Hikaru Sulu (George Takei), Russian navigator Pavel Chekov (Walter Koenig)—and, of course, Leonard Nimoy's star turn as half-Vulcan Commander Spock. Native American first officer Chakotay (Robert Beltran) emerged in the series *Voyager* (1995–2001). The gender balance tended to the heavily male until the advent of *Voyager* Captain Kathryn Janeway (Kate Mulgrew), with half-Klingon chief engineer B'Elanna Torres (Hispanic actress Roxann Dawson). Real-world impacts abound. Nichols, for instance, has related how U. S. civil-rights leader Martin Luther King urged her to remain in the series when she was considering other professional options. Her character, in turn, inspired astronaut Mae Jemison, the first African American woman to be sent into space by NASA.

Fifty years later, how does our world compare with Rodden-berry's universe? The changes in technology are transformational; and although interstellar travel has yet to become reality, NASA's projected 2030s human mission to Mars follows the dream "to boldly go." The progressive social values that Star Trek pioneered on television are now much more widely held. But new conflicts and geopolitical stand-offs have erupted, despite efforts by our own federation, the United Nations. Amid these shifts and tensions, this vastly influential franchise continues to carry a subtle but clear message—we can be better than we are.

References

1. Alcubierre, M. *Class. Quantum Grav.* **11**, L73; 1994.
2. Schurig, D., et al. *Science* **314**, 977–980; 2006.
3. Ni, X., et al. *Science* **349**, 1310–1314; 2015.
4. Takesue, H., et al. *Optica* **2**, 832–835; 2015.

Abstract Theory Has Real Consequences, in the Past and Today

Note: The films discussed here can be viewed at www.labocine.com.

Other than pure math, theoretical physics is the most abstract science and may be even harder to grasp. Pure math has strange symbols and equations galore, but so does theoretical physics. Besides, quantum and relativity theory both ask you to believe in things that make no sense in the world we know, yet are true—that quantum particles pop in and out of existence, that a person traveling near the speed of light hardly ages compared to a twin back on Earth, that mass and energy can be changed into each other.

Creating and understanding these abstractions is not easy, which is why minds like Albert Einstein's are uncommon. But several films in the Labocine collection help by adding visual and dramatic heft to the abstractions or by seeing them from new angles, and by showing how they can change us, especially the theory of relativity through its role in nuclear weaponry.

In *Les lumières ne seront plus jamais aussi rapide* (*Light will never be that fast again,* 2013), a young woman and her friends meditate on two physics mysteries, the speed of light and the nature of turbulence, while gazing at overhead TV monitors on a train platform and later riding on a train. Their thoughts are expressed as a woman in voice-over (the same one we see?) hints at hidden physics enigmas in an urgent whisper (in French with sub-titles) while an eerie soundtrack and deeply shadowed *noir*-like camerawork in black and white intensify the sense of untold mysteries.

The narrator begins by saying "Tragedy began gnawing at me" as she reflects on the speed of light. When the French physicist Hippolyte Fizeau measured that speed in the mid-19th century, he got a value higher than the modern accepted one. The speed of light "has been decaying," our narrator believes; "It's not only melancholia but fear. How low will it go? And if it was to stop? What would we do? Do you have plans for sunset?" Not all her thoughts are so dire. She puts in a vote for French *joie de vivre* when she observes, "Fizeau. . .sounds so French, almost sparkling," practically the sound of frothing champagne; but Michelson and Morley—the Americans who measured the speed of light in 1887 and showed that there is no ether—"sounds like a real estate company to me or maybe a wheel factory."

The film multiply alludes to the theory of relativity. The speed of light is central to the theory, and Einstein himself and many others including myself have used moving trains to demonstrate how the theory changes notions of space and time. There is also a subtle but deep cinematic connection. In its use of voice-over with an unnamed narrator, and unnamed characters; in its unexpected juxtapositions of word and image; in its linkages of emotional states with physical facts; and in its black and white format, this Science New Wave film is highly reminiscent of a great French New Wave film, Alain Resnais' *Hiroshima Mon Amour* (1959).

That classic is the story of a brief love affair between a Frenchwoman and a Japanese man set against the backdrop of the devastation of Hiroshima on August 6, 1945, by an atomic bomb. The bomb converted matter into destructive energy according to the equation $E = mc^2$, which comes from the theory of relativity. Even more significant, the blast would not have happened without Einstein's early intervention.

The story leading up to Einstein's role is told in *Breaking the Chain* (2009), a fictionalized version of real events in 1939, a year after German researchers discovered that uranium nuclei could fission into smaller pieces and release torrents of energy. The Hungarian-born American physicist Leo Szilard worried that the Nazis would produce atomic weapons that created enormous destruction through a chain reaction. He wrote a letter to U. S. President Franklin D. Roosevelt urging the United States to start building "extremely powerful bombs of a new type." Knowing that it would be difficult to get the President's attention, Szilard asked Einstein, by then world famous for relativity, to sign the letter.

That worked, leading to the Manhattan Project that built the bombs the United States dropped on Japan, and eventually to films like *Breaking the Chain*. The film explains the basic physics and shows the initial difficulties: Szilard cannot find a mere $2,000 for his pioneering work on chain reactions (the Manhattan Project eventually cost $27 billion in 2016 dollars); the Nobel Laureate

Enrico Fermi is skeptical about chain reactions but comes around; and the scientists are torn between traditional scientific openness and the extreme secrecy of military research.

All this sets the stage for the Szilard—Einstein letter and what followed, ending in mushroom clouds over Hiroshima and Nagasaki. Few people have seen either cloud close-up and survived, but *As Soon As Weather Will Permit* (2015) by Canadian filmmaker Su Rynard presents the experience of one of them. Rynard's uncle Vernon Rowley was the radar observer aboard *Jabit III*, one of the B-29 aircraft that accompanied the *Enola Gay* when it dropped the bomb on Hiroshima.

The film is based on a letter Rynard wrote to her Uncle Vern and gives his replies in voice-over and writing. These are illustrated with a powerful mix of family and historical scenes along with animation footage, images of B-29s in flight, and the mushroom cloud itself. Rowley takes us through his training for a secret military project, the revelation that the United States will drop an atomic bomb on Hiroshima, and his reactions as the mission proceeds and he sees the actual blast. Rowley's calm manner of speaking brings humanity and ordinariness to a surrealistic and historical event and he simply says, "When you are in the thing, and it's the first one, you don't reflect [on] what it's gonna be, you don't think about it until history has played on it."

Later, summing up his involvement, he writes "We all did what our country asked to the best of our ability, no questions asked," but then suggests that it may not have been all that simple when he says "I didn't have the actual contact devastation. I had the devastation knowing what we'd done, like what have I been a part of." It doesn't emerge in the film but a Google search of Rowley's life and obituary (he died in late 2016) shows that he was a devout Christian. We can wonder how, if at all, his experience with the bomb affected his religious belief.

The history of the atomic bomb began 78 years ago and it has been 72 years since its first and (so far) last use, yet its legacy continues to this very moment. Near the end of her film, Su Reynard tells us, "Vern would say the bomb ended the war. But for me it wasn't an end. It was just the beginning," and she is right. The atomic bomb and its successor, the hydrogen bomb, carried on intercontinental ballistic missiles, dominated the post-World War II era of the Cold War between the United States and the Soviet Union.

That has passed, but now we can worry daily that new mushroom clouds will erupt should a rogue nuclear state like North Korea act as aggressively as it speaks, or if we, the United States, fail to find the right non-nuclear way to deal with it. We cannot yet forget the lesson of the three films I've reviewed: that abstract theory and its practitioners can become pathways to "contact devastation. . .[and] the devastation knowing what we'd done," as Vernon understood out of his own past.

Books about science for general readers by Sidney Perkowitz

Empire of Light: A History in Science and Art (Henry Holt, New York, 1996). Illustrated, paperback and foreign editions, 1998–2005.

Universal Foam: From Cappuccino to the Cosmos (Walker, New York, 2000). Paperback and foreign editions, 2001–2008.

Digital People (JHP/National Academies Press, Washington, DC, 2004; paperback, 2004). Foreign edition, 2011.

Hollywood Science: Movies, Science, and the End of the World. (Columbia University Press, New York, 2007). Paperback, e-book and foreign editions, 2008–2010.

Slow Light: Invisibility, Teleportation, and Other Mysteries of Light (Imperial College Press, London, 2011). Foreign edition, 2014.

Hollywood Chemistry, D. Nelson, K. Grazier, J. Paglia, and S. Perkowitz, eds. (ACS Books/Oxford University Press, Washington, DC, 2013; 2014).

Universal Foam 2.0: From Cappuccino to the Cosmos (Kindle e-book, 2015).

Frankenstein: How a Monster Became an Icon, the Science and Enduring Allure of Mary Shelley's Creation, Sidney Perkowitz and Eddy Von Mueller, eds. (Pegasus Books, New York, 2018).

Physics: A Very Short Introduction (Oxford University Press, Oxford, 2019).

Republishing credits

I'm grateful to the following publications and organizations that granted permission for me to republish my works originally published by them, as listed below. I am the sole author of each piece.

Sidney Perkowitz

Aeon

"Quantum Gravity," *Aeon*, Nov. 11, 2014.
"Light Dawns," *Aeon*, Sept. 18, 2015.
"Crimes of the Future," *Aeon*, Oct. 27, 2016.
"Can a Physics of Panic Explain the Motions of the Crowd?," *Aeon*, Nov. 28, 2018.

Alumni Association, Massachusetts Institute of Technology

"Hubs, Struts, and Aesthetics," *MIT Technology Review*, November/December 1996, 56–63.

American Chemical Society

"Hollywood Science: Good for Hollywood, Bad for Science?," in *Hollywood Chemistry,* D. Nelson, K. Grazier, J. Paglia, and S. Perkowitz, eds. (ACS Books/Oxford University Press, Washington, DC, e-version, 2013; 2014), 263–277.

Association of Alumni and Alumnae of the Massachusetts Institute of Technology

"Brother, Can You Spare a Cyclotron," *MIT Technology Review*, Aug./Sept. 1997, 45–50.

Astronomical Society of the Pacific

"Inspirational Realism: Chesley Bonestell and Astronomical Art," in *Engaging the Heavens: Inspiration of Astronomical Phenomena V, Adler Planetarium, Chicago, Illinois, USA 26 June – 1 July 2005*, ASP Conference Series, Volume 468, Marvin Bolt and Stephen Case, eds. (Astronomical Society of the Pacific, Orem, UT, 2012), 57–62.

Atlanta Magazine

"These Georgia Tech Physicists Helped Prove Einstein Right," *Atlanta Magazine*, Sept. 19, 2016.

¿Cómo ves?

"Art, Physics and Revolution" ("Arte, física y revolución"), *¿Cómo ves?*, Núm. 182, UNAM, Cd. de México, 2014.

"How Realistic are Movies set in Space?" ("¿Qué tan realista es el cine del espacio?"), *¿Cómo ves?*, Núm. 184, UNAM, Cd. de México, 2014.

"Turing and Hawking, Typical Nerds" ("Turing y Hawking, ¿típicos nerds?"), *¿Cómo ves?*, Núm. 197, UNAM, Cd. de México, 2015.

"Brain Injuries in Soccer" ("Las lesiones cerebrales en el futbol"), *¿Cómo ves?*, Núm. 211, UNAM, Cd. de México, 2016.

Creative Loafing

"John Markoff's Love for 'Machines'," *Creative Loafing*, Sept. 3–9 2015, 23.

Denver Quarterly

"Art Upsets, Science Reassures," *Denver Quarterly*, Winter 1996, 120–131.

Discover

"Everything Worth Knowing About...Ice," *Discover*, July/August 2017, 66–69.

Emory Medicine Magazine

"When Vision Betrays: Cataracts," *Emory Health Digest*, Autumn 2017, 20–25.

Evolve

"Future Meat," *Evolve*, June 2014.

Interalia

"Mr. Turner, Artist, Meets Mrs. Somerville, Scientist," *Interalia*, July 2016.

International Society for the Arts, Sciences and Technology (ISAST)

"The Six Elements: Visions of a Complex Universe," *Leonardo*, **43**:2 (April, 2010), 208–211. © 2010 by the International Society for the Arts, Sciences and Technology (ISAST), published by the MIT Press.

JSTOR Daily

"The Internet of Things: Totally New and a Hundred Years Old," *JSTOR Daily*, June 10, 2015.

"The Internet Before the Internet: Paul Otlet's Mundaneum," *JSTOR Daily*, March 5, 2016.

"How to Understand the Resurgence of Eugenics," *JSTOR Daily*, April 5, 2017.

"Do We Have Moral Obligations to Robots?," *JSTOR Daily*, Nov. 29, 2017.

Labocine Spotlight

"Nobody Knows the Quantum," *Labocine Spotlight*, Oct. 19, 2017.

"Abstract Theory Has Real Consequences, in the Past and Today," *Labocine Spotlight*, Oct. 5, 2017.

Los Angeles Review of Books

"Removing Humans from the AI Loop: Should We Panic?," *Los Angeles Review of Books*, Feb. 18, 2016.

"Frankenstein Turns 200 and Becomes Required Reading for Scientists," *Los Angeles Review of Books*, July 9, 2018.

Nature

"Boldly Going for 50 Years," *Nature* **537**, Sept. 8, 2016, 165–166.

Nautilus

"The Case Against an Autonomous Military," *Nautilus Blog: Facts So Romantic*, April 10, 2018.

New Scientist

"Real Physicists Don't Wear Ties," *New Scientist*, Dec. 21/28 1991, 22–24. (c) 1991 New Scientist Ltd. All rights reserved. Distributed by Tribune Content Agency.

"Laughing by Numbers," *New Scientist*, Oct. 12, 1991, 62. (c) 1991 New Scientist Ltd. All rights reserved. Distributed by Tribune Content Agency.

"Spelling it Right in Karachi," *New Scientist*, January 23, 1993, 46. (c) 1993 New Scientist Ltd. All rights reserved. Distributed by Tribune Content Agency.

"In Salmon do did Mobile Bond…," *New Scientist*, 19/26 Dec. 1998–2 Jan. 1999, 62–63. (c) 1998 New Scientist Ltd. All rights reserved. Distributed by Tribune Content Agency.

Physics World

"From Ray Guns to Blu-Ray," *Physics World*, May 2010, 16–20.

"Ad Astra! To the Stars!," *Physics World*, January 2012, 28–32.

"Time Examined and Time Experienced," *Physics World*, July 2018, 20–23.

Science

"The Seductive Melody of the Strings," *Science*, 11 June 1999, 1780.

The American Prospect

"Connecting with E. M. Forster," *The American Prospect*, May–June 1996, 86–89.

The Conversation

"How Close Are We to Actually Becoming Martians?," *The Conversation*, Oct. 1, 2015.

The Science and Entertainment Exchange, National Academy of Sciences

"Science Fiction Covers the Universe and Also Our Own Little Globe," *Science and Entertainment Exchange* (July 2009).

"Food for (Future) Thought or *Star Trek: the Menu*," *Science and Entertainment Exchange* (Feb. 2011).

"Fantasy into science: Invisibility," *Science and Entertainment Exchange* (June 2011).

"Fantasy into science: Teleportation," *Science and Entertainment Exchange* (Sept. 2011).

"Fantasy into science: Tractor beams," *Science and Entertainment Exchange* (Dec. 14, 2011).

The Sciences

"True Colors," *The Sciences*, May/June 1991, 22–28.

"Strange Devices," *The Sciences*, January/February 1995, 21–27.

"Stealth Science," *The Sciences*, November/December 1995, 40–44.

"Froth with Meaning," *The Sciences*, March/April 2000, 34–38.

Vitra Design Museum

"Illuminating Light," in *Lightopia,* M. Kries and J. Kugler, eds. (Vitra Design Museum, Weil-am-Rhein, Germany, 2013), **1**, 13–26.

Index